A Catholic and Hands-on Appro

Behold and See 5

David Beresford, Ph.D.

Catholic Heritage Curricula

1-800-490-7713   www.chcweb.com
P.O. Box 579090, Modesto, CA 95357

To Theresa, my dear wife, who made everything possible;

and to my children, Philippa, Hugh, Elizabeth, Anna,

Patrick, Sally, and Margaret, and Anthony (who is with

Our Lord and His Blessed Mother to pray for us)

ISBN: 978-0-9824585-4-9

© 2010 Theresa A. Johnson

This book is under copyright. All rights reserved. No part of this book may be reproduced in any form by any means—electronic, mechanical, or graphic—without prior written permission. Thank you for honoring copyright law.

Image Credits may be found on page 226.

Distributed by:
Catholic Heritage Curricula
P.O. Box 579090
Modesto, CA 95357
www.chcweb.com

Printed by Sheridan Books, Inc.
Chelsea, Michigan
April 2015
Print code: 371956

⚠ Please note: Although every effort has been made to ensure the safety of all experiments within this program, users are responsible for taking appropriate safety measures and supervising children during experiments. Catholic Heritage Curricula disclaims all responsibility for any injury or risk which is incurred as a result of the use of any of the material in this program.

Contents

Introduction, *iv*

Chapter 1: Studying Living Things around Us, *1*

Chapter 2: Metamorphosis, Creatures Helping Other Creatures, and Graphing Our Findings, *19*

Chapter 3: Food Webs, Resistance to Disease, and Conservation of Energy, *39*

Chapter 4: Physiology and Introduction to Biochemistry, *69*

Chapter 5: The Circulatory System and Human Physiology, *85*

Chapter 6: Logic: Deduction, Induction, and Scientific Reasoning, *97*

Chapter 7: Competition among Plants and Animals, *113*

Chapter 8: Atmosphere and Weather, *137*

Chapter 9: The Earth and Its Composition, *173*

Chapter 10: Genetics and Taxonomy, *195*

Indexed Glossary, *220*

Supply List, *225*

Question the beauty of the earth . . .
question the order of the stars . . .
question the living creatures that move in the waters,
that roam upon the earth,
that fly through the air . . .
question all these.
They will answer you:
'Behold and see, we are beautiful.'
Their beauty is their confession of God.
Who made these beautiful changing things,
if not One Who is beautiful and changeth not?
—St. Augustine

Behold and See 5

Student Text: A detailed supply list for the experiments in this text can be found on page 225. An indexed glossary is also provided in the back of the text for students' use. Answers to questions posed in the text are included in the back of the student workbook, along with answers to the workbook exercises and tests.

Student Workbook: The student workbook provides kid-friendly exercises and tests to accompany the text. Encourage your student to refer to the text to complete the workbook pages.

Daily Lesson Plans: A detailed daily lesson plan to assist in teaching this course can be found in *CHC Lesson Plans for Fifth Grade* or in *Behold and See 5 Daily Lesson Plans,* a single-subject lesson plan that coordinates reading assignments, exercises, experiments, supply lists, and tests in a clear, attractive format. Visit *www.chcweb.com* for more information.

A Catholic and Hands-On Approach to Science

Firmly grounded in the belief that faith and reason are inseparable, the science materials published by Catholic Heritage Curricula teach up-to-date scientific knowledge within the context of our Catholic faith. Scientifically excellent, these texts allow students to progress rapidly in their understanding of scientific discoveries and the scientific method. More importantly, they gently but compellingly demonstrate God's active and foundational role in creation, and reflect on the proper use of scientific knowledge for the glory of God. Learn more about the science programs offered by Catholic Heritage Curricula by visiting *www.chcweb.com*, or request a free catalog by calling 1-800-490-7713.

Dear Parents and Young Scientists:

We know that God created us, the world, and everything in it. This creation is profound, for God not only created things, He created the very possibility for anything to exist. God made space for things to stretch into, and time for things to exist in, and most fundamental of all, he created existence itself, from nothing! God continues His creation by holding everything in existence from moment to moment. This may seem an odd way to look at it, but if God stopped thinking about something, it would return to its original state of nothingness and cease to exist! But we know this will never happen, for nothing is impossible for God.

Of course, because everything comes from God, everything can teach us about God if we see things in right order and with the eyes of faith. But if we do not know God, or look with the eyes of faith, we can get caught up with the beauty of creation and miss the Creator.

Because of the relationship that God has with His world, everything that we can study in science can bring us closer to a knowledge of God. And the best way to keep close to God when studying science is by staying close to His Son Our Lord.

Mary, Mirror of Justice, Seat of Wisdom, Cause of our Joy, pray for us.

—David Beresford, Ph.D.

David Beresford, Ph.D. (population biology/entomology), teaches biology at a variety of levels: university courses in Introductory Biology, Statistics, Euclidean Geometry, and Mathematics at Our Lady Seat of Wisdom Academy, and Entomology and Invasive Species Biology at other institutions; high school biology at Wayside Academy, a private Catholic school, and elementary science to his own children at home. He writes articles about G.K. Chesterton for *Gilbert! Magazine* and has also been published in *Catholic Insight* and other magazines. His scientific research focuses on finding ways to control the insect pests of dairy and beef cattle and studying species' diversity in northern habitats. This work has been published in various ecology and entomology scientific journals. He lives with his wife and children on a farm in Dummer Township, Ontario, Canada, where they raise pigs, geese, and chickens.

1
Chapter

Studying Living Things around Us

[St. Thomas] was willing to begin to study the reality of the world in the reality of the worm. His Aristotelianism simply meant that the study of the humblest fact will lead to the study of the highest truth.
— G.K. Chesterton

Studying Living Things around Us

Every living thing was created for a purpose and has its own place in God's creation.

Words to Know

Biology is the study of living things.

Zoology is the study of animals.

Botany is the study of plants.

This year you will learn about many different living things. Just as important, you will learn *how* to study in a scientific way. Science can be learned from textbooks, but it is first learned by paying attention to, or observing, the living things that we want to learn about.

God has created so many things to study in science that there are different fields of study for almost every created thing. People who study rocks are called geologists; people who study volcanoes are called volcanologists. There are even people called mycologists, who study mushrooms!

Your studies this year will focus on living things and the Earth they inhabit. We will begin with the study of living things, or **biology**. Biology can be divided into many categories, including the study of animals (**zoology**) and plants (**botany**). Each of these branches of study can be divided again into the study of specific animals or plants.

For example, biologists who study animals are called zoologists, and the subject of zoology can be subdivided as:

Type of Study	Subject Name	Person Who Studies This
fish	ichthyology	ichthyologist
birds	ornithology	ornithologist
mammals	mammology	mammologist
reptiles and amphibians	herpetology	herpetologists
insects (and spiders and mites)	entomology	entomologists
insects that affect health	medical entomology	medical entomologist
how living things fit together	ecology	ecologist
how populations grow	population biology	population biologist

Biologists who study plants are called botanists, and the subject of botany can be subdivided as:

Type of Study	Subject Name	Person Who Studies This
crop production	agronomy	agronomist
forest management	forestry	forester
herbs and their uses	herbology	herbologist
cultivated plants	horticulture	horticulturist
plants that live in the ocean	marine botany	marine botanist
ancient fossil plants	paleobotany	paleobotanist
plant diseases	phytopathology	phytopathologist

A scientist can be one or many of these things. I study flies that attack cattle, so I am a veterinary entomologist. I can also be called a dipterist (a person who studies flies), a population biologist, and an ecologist. Which of these types of studies is interesting to you?

You are going to look at some animals and plants. They are alive, just like we are, and they share the world with us. We are able to learn about them, and the more we learn, the more we are grateful that they are here with us! Every living thing enjoys being alive and being here with us on earth.

Earthworms

There are no ugly animals, or ugly plants! But some have jobs to do that we would not enjoy doing. One example of this is earthworms. Earthworms spend all day eating soil and sand, and any rotten plants or dead insects that are in that soil and sand. They are shaped like blind living tubes, and they eat their way through the soil. Farmers need earthworms to help grow crops, so worms are helpful to us. But most important, we need earthworms because they teach us about the world we share. And if you look closely, earthworms are beautiful! Look at the shiny skin, the perfect shape for living in the soil, and the subtle colors. And think! Where they live, in the dark soil, their colors are not seen; they cannot even see each other because they have no eyes. Yet, God gave them their own beauty, too.

Body Plan

A worm's general shape is a tube inside a bigger tube. The inside tube is its gut, or stomach. At the front end of this tube, sand, soil, and food goes in through the mouth. At the back end of this tube, sand, soil, unused food, and waste come out of the anus as a very rich food for plants. The outside tube is made up of the worm's skin. Between these tubes is muscle which it uses for moving. The worm pushes its mouth into the soil, and works its way through the soil eating what is in front of it, leaving bundles of rich plant food behind.

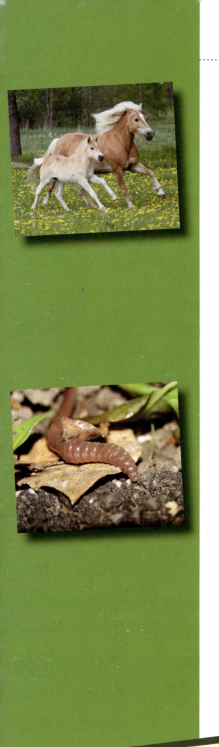

Earthworms lay eggs called cocoons in the soil. When the eggs hatch they are small, about 1/8 inch long or smaller.

Number of Species

There are 33 species of worms in the United States and 19 species in Canada. A **species** is a group of animals or plants that are all of the same kind, like tulips, robins, or house flies. There are probably more worm species in North America that we do not know about yet. New worms and other species of animals and plants are always being discovered. Would you like to be a scientist who discovers new species?

European Origins of Earthworms in North America

Almost all the worms in the United States and Canada are originally from Europe. Here is why.

About 2 million years ago, the climate on Earth was cooling down. Because of this, snow piled up and did not melt in the spring. As snow kept piling up, it got compacted into thick sheets of ice called **glaciers**.

When glaciers covered the northern parts of North America, they killed all the earthworms under them. This is not surprising. Imagine being a worm, or anything, under a big block of ice that is 10,000 feet high! These large ice sheets covered almost all of the northern half of the United States and all of Canada and Alaska. Snow kept falling and piling up into huge, heavy mounds of ice. The ice flowed outward from the center (ice can flow, but very slowly!) toward the south. Ice covered so much of the continents that this period of time is called the **Ice Age**.

The moving ice pushed away all the soil and rocks underneath it, like a large snow plow. You can see where the ice pushed the

Words to Know

A **species** is a group of animals or plants that are all of the same kind.

A **glacier** is an extremely thick sheet of ice.

The **Ice Age** was a period of time when ice covered much of the continents.

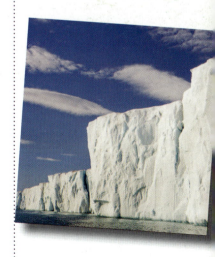

soil away if you travel north towards the Great Lakes, past New England, to Canada. There, you will find that instead of soil, the ground is a huge rock of pink, red, black, and white granite. This is called the Canadian Shield, because it covers most of Canada. Sometimes there is moss growing on it, and where the thin soil has been washed by the rain, trees grow out of the cracks in the rocks. This is only twenty miles north of where I live, and I can drive there and stand on the rocks that were scraped clean by glaciers.

When the ice melted, it left the soil and rocks all along its edge. That is why the soil is so thick in the corn belt area of the Midwestern United States. It is also why there are so many rocks in New England and Ontario. At my farm near the Great Lakes, the ground is filled with grey limestone rocks. It is impossible to dig a hole without hitting these rocks.

The ice melted from the middle parts of North America 25,000 years ago. If you want to see glaciers today, you need to go to Alaska, northern Canada, or Greenland. If you go to these glaciers, you will have to be careful, because snow is still falling on them, and sometimes the snow slides off, knocking down all the trees and anything else that is in the way. This is called an avalanche.

How European Earthworms Came to North America

After the ice melted, and the lakes and rivers filled with water, there were no earthworms where the ice used to be. Earthworms move very slowly, and usually get from one place to the other by being carried by people. When the first settlers came from Europe to North America, bringing with them new plants and seeds for their gardens, worms hitched a ride. These settlers also brought apples, cows, pigs, sheep, beans, lettuce, spinach, and other crops. Also among the seeds, and in the soil, they brought weeds, insect pests, caterpillars, and lots of earthworms. As more and more of the forests were cleared for farmland, earthworms spread throughout the entire continent.

Earthworms and Farms

Without earthworms, there would not be productive farms, grain, corn, vegetables, beef, chicken, or pork, because cows and chickens and pigs feed on grass and grain grown from worm-rich soil. Worms keep adding nutrition to the soil, and help repair the soil when it becomes damaged from trying to harvest too much from it.

Earthworms dig tunnels and, because of this, they bring new soil up from below to where the plants are. The tunnels allow air to get into the soil to keep it friable, which just means loose and nutritious. Without this topsoil, nobody could live on earth. And without earthworms, it would be very hard to feed everyone, because worms bring lots of nutrition to our farm crops. When farmers put manure on the soil to feed the crops, earthworms spread the manure around so all the crops can grow.

Earthworms and Crop disease

You remember that God had a reason for everything that He created? Well, earthworms have another important job. They help prevent disease from spreading to apples! Earthworms like apple leaves that fall from the trees in the autumn. They grab these leaves and pull them into their holes. There is a kind of **fungus** called apple scab that attacks apple leaves and fruit. If the leaves are left on the ground, the fungus **spores** go through the air and land on the new leaves each spring, harming the trees and damaging the apples. Because earthworms pull the old

Words to Know

A **fungus** is a type of plant that is often harmful to other living things.

Spores are similar to tiny, airborne seeds.

Words to Know

Pesticides are chemicals used to kill insect pests.

A **habitat** is a place or environment where something lives.

Classification is a scientific system that groups things into classes with other like things.

Taxonomy is the scientific name for classification.

---

leaves underground, the fungus cannot attack the new leaves in the spring. Apple orchards with lots of earthworms do not have much problem from scab and other fungi.

Earthworms help farmers. But sometimes farmers have to spray **pesticides** (chemicals that kill insect pests) to kill insects that attack their crops. When the rain washes this spray into the soil, it kills the earthworms. Then, just as the farmer thinks that he has protected his fruit trees from insect pests, apple scab and other fungi attack the trees the next year. Why? Because, along with the harmful insects, the pesticides also killed some of the helpful worms.

This is often how scientists learn how important things like earthworms are, by accidentally killing them when trying to solve some other problem. Because of this, agricultural scientists are always looking for new pesticide sprays and methods to protect our fruit crops from insect pests without harming earthworms or other helpful insects.

Kinds of Earthworms and Habitats

There are different kinds of earthworms, and they live in different **habitats**. (A habitat is a place or environment where something lives.) Big night crawlers (6 inches) live in the lawn as do smaller earthworms (3 inches). There are green and brown manure worms (you can guess where they live); worms that live in rotten logs and even plugged eaves' troughs around your house; and very small red wrigglers (2 inches long). Some of these worms live in the woods, some beside streams and ponds, at farms, in forests, and under rocks and logs. Maybe you have seen some of them when you were camping or fishing or playing outdoors.

Classification

In God's creation, everything has a place where it fits in relation to everything else. About three hundred years ago, a man named Carl von Linne, or in Latin, Linnaeus (1707 to 1778), realized that it would be easier to study plants and animals with a system that showed where they fit in God's creation. So he developed a system, called **classification**, that placed everything into its own group or class.

Procyon lotor

This system of classification is called **taxonomy**. To keep track of where animals and plants fit within this grouping, species were identified by using two names: the genus name and species name. Linnaeus gave us many of the classifications we still use today. Since the time of Linnaeus, many more levels and divisions have been added to classification charts but, to give you an idea, think of sorting everything in your house into groups of similar things.

Orconectes rustica

For example, all furniture could be grouped together, and all clothing, and all cooking and serving utensils. If you wanted to "classify" where a silver baby spoon fit, you could make a chart something like this:

Vulpes fulva

Kingdom:	Household (factories or businesses would be different Kingdoms)
Phylum:	Kitchen
Class:	Food preparation or serving
Order:	Non-sharp eating utensil
Family:	Spoon
Genus:	Silver spoon
Species:	Baby spoon

Do you see how a Kingdom includes a huge group of all kinds of things, but as we move from Phylum to Class to Order and on down the line, more and more things are eliminated until we get down to smaller and smaller groups of things, animals, or plants?

The groups go from largest to smallest as: Kingdom, Phylum, Class, Order, Family, Genus, Species.

	raccoon	red fox	crow	house fly	rusty crayfish	sweet pea tree
Kingdom	Animalia	Animalia	Animalia	Animalia	Animalia	Plantae
Phylum	Vertebrata	Vertebrata	Vertebrata	Arthropoda	Arthropoda	Anthophyta
Class	Mammalia	Mammalia	Aves	Insecta	Crustacea	Angiospermae
Order	Carnivora	Carnivora	Passeriformes	Diptera	Decapoda	Rosales
Family	Procyonidae	Canidae	Corvidae	Muscidae	Astacidae	Leguminosae
Genus	*Procyon*	*Vulpes*	*Corvus*	*Musca*	*Orconectes*	*Lathyrus*
Species	*lotor*	*fulva*	*brachyrhynchos*	*domestica*	*rustica*	*odoratus*

Words to Know

Invertebrates are animals that have no backbones.

Vertebrates are animals that have backbones.

Notice that the mammals and birds are in the same Phylum, as are the fly and crayfish. The raccoon and red fox are together until Order Carnivora when they separate into different Families. The complete name is, for example: rusty crayfish, *Orconectes rusticus*. The Latin name must be either italicized or underlined (but not both!), so you could write it as Orconectes rusticus. For the red fox, it is *Vulpes fulva* or Vulpes fulva, but never Vulpes Fulva, Vulpes fulva or vulpes fulva.

Earthworms belong to the Animal Kingdom (Kingdom Animalia), and are **invertebrates,** meaning they have no backbones. (People have backbones, so they are **vertebrates**.) Earthworms belong to the Phylum Annelida (segmented animals), the Class Oligochaeta (few bristles), and the Order Haplotaxida. The Common Earthworm belongs to the Family Lumbricidae, and its scientific name is *Lumbricus terrestris* (earthworm), which tells us its genus and species. Earthworms that live in soil are also called megadriles.

Predators, Parasites, and Worms

Worms have their enemies, too. Lots of animals are **predators**, or animals that seek out and eat other animals. (Think of a hawk catching a mouse. Hawks are predators.) Predators such as fish, birds, moles, shrews, and insects all enjoy a good meal of worms. In this case, the worms are **prey,** or things that are hunted and eaten by the predators.

When some birds eat earthworms, the worms sometimes make the birds sick. This is because other tiny animals live in the worms, and hitch a ride on them, causing disease in the birds that eat them.

Animals and plants that live on, or feed off of, other living things are called **parasites**. Some parasites bite and suck blood from the animals that they live on, like fleas on cats and dogs. Other parasites are helpful, like certain bacteria on your skin which can protect it from disease.

There are also deadly parasites. One of the deadliest enemies of earthworms is the cluster fly. These are the flies that come into houses every autumn looking for a warm place to spend the winter. Cluster flies are a kind of blow fly that lays eggs on earthworms or on the soil where earthworms live. The fly eggs hatch into **larvae**, which also look sort of like worms. These fly larvae burrow into the earthworm to eat it, eventually killing it. Then each fly larva forms a **pupa**, a sort of sleeping-bag-like cocoon, where it finishes growing into an adult fly. Every fall there are thousands of flies emerging from lawns and gardens; these get into our houses. They do not do any damage to people, other than irritating us by buzzing around the lights and windows. But the cluster fly is one of the earthworms' worst enemies.

Even lowly earthworms and cluster flies have a purpose in our world. Every living thing was created for a purpose and has its own place in God's creation.

Words to Know

Predators are animals that hunt for and eat other animals.

Prey are creatures that are hunted and eaten by predators.

Parasites live on, or feed off of, other living things.

A **larva** is the worm-like, immature form of an insect.

A **pupa** is a cocoon in which immature insects grow into adults.

> Let's put what we've learned to work.

Roll up Your Sleeves!

Behold and See 5 Student Workbook:

Worksheets for Chapter 1 begin on page 1.

Please remember to read all of the activities, right to the end of the chapter, even if you are assigned only one or two activities. Information contained in the activities is also instructional, and part of your lessons!

First:

You will need a bound notebook with blank pages, a pencil, and colored pencils. Put your name on the front of your notebook. Also include the month, year, and where you live. You will be keeping notes in this book, and drawing pictures and diagrams. A notebook is the most important tool of science and every scientist has a notebook. I still have all my old ones, and they are the most important thing I own because they have all my research in them!

Do not use a loose-leaf binder for your notebook. The pages get lost, you cannot put the notebook in your field bag or backpack, and it takes up too much room. Use a small notebook, about 5 to 6 inches long. And use pencil rather then pen if you can. Diagrams should be colored, as coloring makes it much easier to identify specimens.

If you have a magnifying glass, use it to look at specimens. NEVER LOOK AT THE SKY WITH A MAGNIFYING GLASS: YOU WILL BE BLINDED!

Lab Activity #1

Get your notebook, a pail, and a shovel. Go into your yard and dig for earthworms. Measure them. How long are they? How wide? In your notebook, draw a picture of them (use a pencil or black pen, and colored pencils). Label the parts of the earthworm and color the worm just the way it looks in real life.

Look at the earthworm. What colors do you see? Notice how the worm is divided into segments. This is why it is called a segmented worm (there are other kinds of worms that are smooth-sided). Now feel the worm. The slimy feeling is due to mucus which protects its skin from bacteria.

Did you notice that the worm was also rough along the side? This is due to lots of small hairs or bristles called setae. The bristles are like cleats on shoes to help grip the sides of their tunnels so they can move.

Do you have any old boards or rocks in your yard? Look under these; earthworms like to hide under rocks and logs. What other creatures did you find when you were looking for earthworms? Draw what you found.

What else did you find? Be careful! You might find red ants or fire ants, and these can sting!

Put a board out on the garden, and water it. Look under this board each day to see what you might find. How many days did it take for earthworms to appear?

You can keep earthworms in a container as pets for study, as long as you feed them. They eat leftover lettuce, potato peelings, and apple peelings. They also need water, so sprinkle the soil with water. Cover the top of the container tightly with netting or cloth, so that the earthworms cannot escape and flies do not get into the worm food. Keep them in a cool place.

Lab Activity #2

Go out in the back yard and collect some earthworms. *(If you cannot find anywhere to dig, you can use my numbers for this study.)* Keep count of how many worms you catch in each shovelful of soil. Write these numbers down. Here is how many I found in my back garden when I dug 12 times: I got nothing in the first shovel, 1 in the second and third, nothing in the next three. Then I got 2 worms in the 7th shovel, then 1 and 4 in shovels 8 and 9, and nothing again for shovels 10, 11, and 12. I wrote this down in my notebook like this:

Log: Earthworm   Place: Warsaw, Ontario
Name: David Beresford   Date: July 9, 2009

Number of worms in each shovel of soil, back garden:

1st shovelful:	0	7th:	2
2nd:	1	8th:	1
3rd:	1	9th:	4
4th:	0	10th:	0
5th:	0	11th:	0
6th:	0	12th:	0

12 shovels of soil, 9 worms caught

draw lines on paper at the end of each worm

Put your earthworms in a pail or large bowl (a salad bowl will work well for this), and put some soil in with them. We are going to measure how long they are. The best way to do this is to gently place them on a paper and mark where the front and back of each worm is. Try to hold them straight, but do not stretch them or they might break in two!

Then, put the worms back in their container. Measure the length of each worm by measuring the distance between the lines with a ruler. Record the data (the lengths) in your notebook. Give each worm a number so you can keep track of it. When I did this, I got these numbers:

worm #	length in inches
1	2 ½
2	2 ¼
3	1 ¾
4	1 ¾
5	1 ¼
6	3 ½
7	3 ¼
8	2 ¾
9	2 ¼

How can we show this information? A lot of numbers is not very helpful. The best scientific way is to draw a graph of the worm lengths. Make sure you put the labels on each axis. An **axis** is the line on the edge of a graph (see green axes below). My graph looks like this:

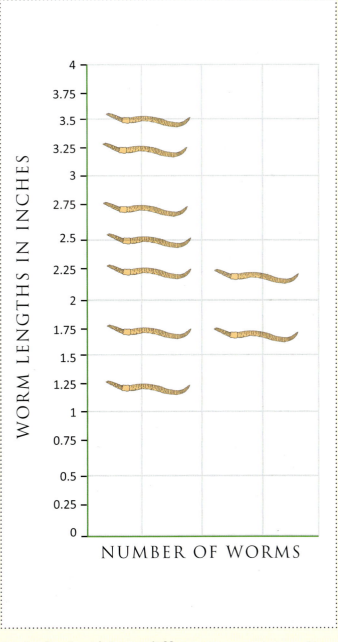

A decimal is a different way to write fractions. ½ is 0.5, ¼ is 0.25, and ¾ is 0.75.

I dug another set of holes, this time from the front lawn. Now I can compare how many worms there were in each place. I use a different kind of graph for this, called a bar graph. First, I make a table to prepare the graph. I list the number of shovelfuls that had any worms in them. The table looks like this:

Table 1. Worms caught by digging, Warsaw, Ontario, July 9, 2009.

	back garden	front lawn
number of shovelfuls	12	12
number of worms caught	9	4
number of shovelfuls with worms	5	3

Now using the numbers in the table, I made a graph, labeled the axes, and put a sentence underneath it to explain the graph.

Do you see what this graph tells us? More shovelfuls had worms in them from my back garden than from my front lawn. We can use the information from the graph to guess that there are probably more worms overall in the back garden than in the front lawn. Then we could try to figure out why this is. Perhaps there is more water in the back garden, or maybe the worms are particularly fond of the compost in the back garden.

Does your back yard have more or fewer worms than your front yard (or a second location)? Find out by digging another set of holes. Use your new data to make a bar graph similar to mine.

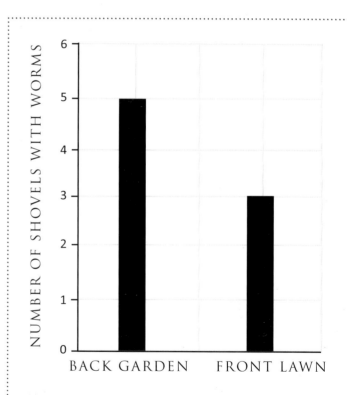

Figure 1. Worms caught in Warsaw, Ontario on July 9, 2009. The number of shovels of soil that had worms in the back garden and front lawn. There were 12 shovels of soil dug at each place.

Activity #3: Writing Activity

Write a short report on earthworms.

Include a **Title**, **Introduction**, **Methods**, **Results**, and **Discussion** section.

For the **Title**, tell what you looked at and where.

In the **Introduction** explain why earthworms are important.

In the **Methods** section, tell what you did (digging with a shovel) and when. Also tell how you measured the earthworms.

For **Results**, tell what you found and include your graphs but not your tables. (The information in the tables is included in the graphs. You should never include the same information twice.)

Make sure you invite the reader to look at your graphs. The best way is to write something like this: "I found more worms in the back garden than the front lawn (Figure 1)."

In the **Discussion**, explain why worms are important for farmers and all of us who like to eat!

Make sure to include a drawing of a worm and some of the other creatures you found in the soil.

Chapter 2

Metamorphosis, Creatures Helping Other Creatures, and Graphing Our Findings

For myself, I confess that I do not know why rats and mice and frogs were created, or flies, or worms; I only see that all such creatures are beautiful in their own kind . . .
— St. Augustine

Metamorphosis, Creatures Helping Other Creatures, and Graphing Our Findings

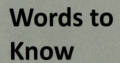

Words to Know

A **grub** is the worm-like, larval stage of an immature beetle.

To **pupate** means to finish growing into an adult insect inside a cocoon-like structure.

Metamorphosis is the process of going through different, changing forms to reach adulthood.

Burying Beetles and Metamorphosis

One thing that we biologists discover very quickly is that many animals are very good parents. We know that birds are good mothers, and so are alligators and crocodiles, but insects can be outstanding parents as well!

I caught some beetles in a trap with some dead mice in it. These beetles are named burying beetles because they bury dead animals by digging a hole underneath them and then covering them with soil. Burying beetles are an excellent example of how even the smallest creatures take a lot of care in raising their children.

Both the mother and father beetle fly around an area looking for a dead mouse or squirrel to bury. They bury the dead animal to hide it from other animals like flies or skunks that might want to eat it. Once a dead mouse is found, the male and female beetles hollow out a small chamber in the soil by the mouse. Then the mother beetle lays her eggs in the chamber. Both the mother and father guard their nest.

The eggs hatch and grow into baby beetles, called **grubs**. The mother beetle brings the grubs to the food (the dead mouse) and then feeds her young grubs one at a time from her own mouth, just like a mother bird does.

As a grub grows, it grows out of its skin! When the grub is ready to molt, or shed its skin, the mother feeds it again. Then, once the grub is finished growing, it **pupates** in the soil, and the parents' work is done. In this resting pupal stage, the young beetle, or pupa, grows inside a cocoon covering rather like a sleeping bag, and then finally pops out as an adult beetle. When the adult beetle emerges, its transformation from egg to grub to pupa to adult is complete. (This process of going through different, changing forms to develop into an adult is called **metamorphosis**.) Now the new adult beetle begins looking for a dead animal of its own so it too can start a family.

Metamorphosis

Growing up surrounded by adorable babies, puppies, and kittens, it is easy to forget that not all of God's creatures start as small versions of adults. "Growing up" means something far different for insects.

Beetles aren't the only insects that go through complete or incomplete metamorphosis to reach adulthood. Dragonflies start their lives as swimmers! And butterflies and moths metamorphose from egg, to larva, to pupa, to adult as well. In fact, it is almost worth raising tomatoes just to see if you can capture an overstuffed, green Tomato Hornworm caterpillar and witness it pupating into an adult moth. You can do the same thing with butterfly caterpillars, the larval stage of butterflies. You can plant pink or purple zinnias (butterflies are attracted to pink and purple), and capture caterpillars for pupating that way.

Perhaps you would like to capture several kinds of caterpillars, keeping record of how long it took each to pupate, as different insects pupate on different timetables. By recording your findings, you become better prepared and more successful at raising critters through their developmental cycles. Do you see how recording your discoveries through notes and graphing can help in learning about the world around us?

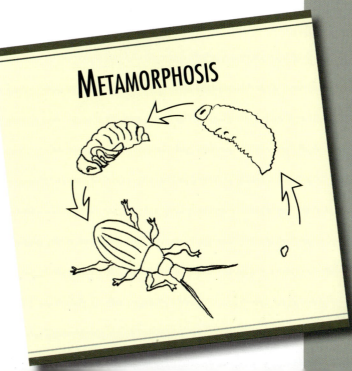

Metamorphosis

21

Words to Know

Facilitation is a working relationship between creatures that helps both creatures.

Phoresy is a working relationship in which one creature "hitches a ride" on another.

Pollination is the process of spreading pollen to flowers to produce fruit and seeds.

Relationships between God's Creatures

With so many animals looking for food, finding something like a dead mouse is a real treasure. Flies also want to find these dead animals so they can lay eggs on them. So, the burying beetles team up with even smaller animals, called mites, to keep the flies from claiming the dead mouse. This working relationship is called **facilitation**, which means that it helps both the beetles and the mites.

This is how the mites and the beetles help each other. The mites climb onto the young adult beetles when they are leaving the soil. Then, when the beetles find a mouse of their own, the mites hop off and eat as many fly eggs as they can find. Then there won't be any fly eggs to hatch and eat the dead mouse, which would take food away from the baby beetles. This way, the mites help the beetles by saving their food from the flies, and the beetles help the mites by giving them a ride.

The mites stay with the nest, laying eggs of their own. When the new beetles leave, young mites climb onto the beetles and hitch a ride, even when the beetles are flying! This hitching a ride is called **phoresy** (pronounced fouressey), and the relationship is called a phoretic relationship.

Burying beetle carrying mites.

A similar relationship is that between bees and flowers. For example, when bees get nectar from flowers, they use it to make honey for food. At the same time, while the bee is gathering nectar from the flower, it pollinates the flower, which is helpful to the flower. (**Pollination** is the spreading of pollen, which is produced by the flowers the bees visit.) As the bees travel from flower to flower, pollen is brushed onto their bodies and carried to new flowers. Without pollination, no fruit or seeds would be produced and no new plants would come from the parent plant. Without bees, flowering plants would eventually die out and, without nectar, bees would also eventually die out. Both the bee and the flower benefit from facilitation, and so do we. What would we do without honey, flowers, fruit, and other foods that grow from seeds? Aren't God's plans for nature, and for us, amazing?

Although plants and animals often help each other, they also compete with each other for food and the other things they need to survive. We have been looking at carrion beetles, so let us use these as an example.

There are 15 different kinds of carrion beetles in the United States and Canada that belong to the genus *Nicrophorus*. Some are best at laying eggs in large birds and dead seagulls (*Nicrophorus investigator*). Some of the beetles are covered in fur that helps keep them warm when they fly (*Nicrophorus tomentosus*). Others, because they are smaller and cannot fight for carcasses very well,

Words to Know

A **parasitoid** is somewhat like a cross between a predator and a parasite.

A **host** is an animal or plant that a parasite lives and feeds on.

have a very good sense of smell and are better at finding places for their eggs (*Nicrophorus defodiens*) before bigger beetles do. One of the strangest kinds of burying beetles lays eggs in live snake eggs instead of dead animals (*Nicrophorus pustulatus*). Because the beetle kills the live snake eggs to lay its own eggs, it is called a **parasitoid**. A parasitoid is halfway between a predator and a parasite.

You remember that predators seek out and kill their food, while other insects and animals are parasites that live right in or on another insect, plant, or animal. These parasites take, or suck, their food right from the **host**—the plant or animal that they are living on—but they do not kill the host. For example, a flea is a parasite that feeds on the blood of the dog, or host, that it lives on.

Table 1 below summarizes the relationships between creatures.

Table 1. Type of relationships, from facilitation to predatory

type of relationship	definition	example
facilitation	helps the host and gets a benefit as well	bees pollinating flowers and getting nectar
phoresy	uses the host for travelling, hitching a ride	mites on burying beetles
parasite	lives on other living things (the host), using them for food, but not killing the host	mosquitoes on people lice on cattle or sheep fleas on dogs, mites on flies
parasitoid	attacks the living host, eventually killing it	burying beetle on snake eggs wasps on caterpillars
predator/prey	kills and eats its host	dragonflies eating mosquitoes hawks eating pigeons

Collecting Data and Presenting Your Results

So far we have looked at how to collect information and how to record it. Now we are going to look at how we tell others about what we find in biology using graphs. Here are some results of a test I did.

A fellow scientist and I were looking at how big the male and females beetles were, and we got the following data for a burying beetle called *Nicrophorus tomentosus* (Table 2):

Table 2. Caught in traps baited with dead mice and birds. Sizes are in millimeters (3 mm is ⅛ inch, and 6 mm is ¼ inch). August, 2008 Algonquin Park, Ontario, Canada

female					male				
16	18	20	21	22	15	19	21	22	24
17	19	20	21	22	18	19	21	22	24
17	19	20	21	22	18	19	21	23	24
18	19	20	21	22	19	20	21	23	24
18	19	20	21	23	19	20	21	23	24
18	19	20	21	23	19	20	22	23	25
18	19	20	21	23	19	20	22	24	26
18	20	20	21		19	20	22	24	
18	20	20	21		19	20	22	24	
18	20	21	21		19	21	22	24	

Female and male burying beetles.

We can make a graph using the data. To do this we need to count how many of each size there are, which we do by making a tally table like the one below. (Tally marks work like this: 1 is |, 2 is ||, 3 is |||, 4 is |||| and 5 is ⊥⊥⊥⊥.)

length in mm	females	males
15			
16			
17				
18	⊥⊥⊥⊥ |||	||
19	⊥⊥⊥⊥		⊥⊥⊥⊥ ⊥⊥⊥⊥
20	⊥⊥⊥⊥ ⊥⊥⊥⊥			⊥⊥⊥⊥	
21	⊥⊥⊥⊥ ⊥⊥⊥⊥		⊥⊥⊥⊥	
22	||||	⊥⊥⊥⊥ ||
23									
24		⊥⊥⊥⊥ ||||
25			
26			

Now, using bars, we can make one graph for each gender, like the graphs on the next page. Each bar represents the number of beetles of a certain size. For instance, eight of the female beetles were 18 mm long, so the bar that is labeled "18" on the horizontal axis is 8 squares high. Bar graphs like these allow us to show how many beetles there were of each length in a way that can be understood at a glance.

Study the bar graphs on the next page and use them to answer the following questions.

- What is the smallest beetle? Is it a male or a female?
- What is the biggest beetle? Is it a male or a female?
- What is the range of sizes for males, and for females? (Remember that the range is just the largest length minus the smallest length.)

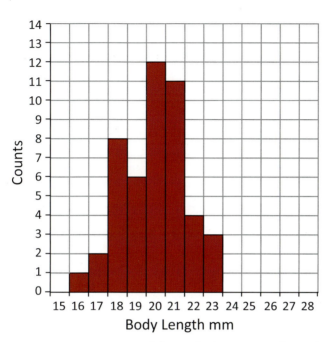

Figure 1. Size of female burrying beetles

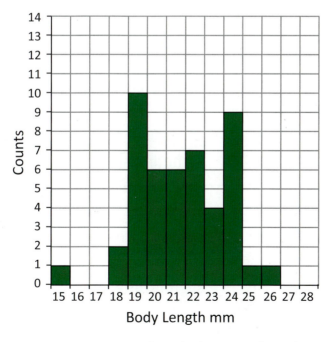

Figure 2. Size of male burying beetles

Variability: Differences Within Species

Notice that the smallest beetle was a male, but overall the males were larger. This is an extremely important point. All living things are slightly different, even in the same family, or same species, or same group. Differences that occur within the same group or species is called **variability**.

Understanding variability, or differences within a group, is the basis of all modern biology. Because of variability, farmers and animal breeders have made hundreds of different kinds of apples, dogs, sheep, cattle, horses, potatoes, corn, and all kinds of things, by cross breeding for **characteristics,** or special differences, in each plant or animal. These differences make the plants and animals more useful for specific purposes.

For example, some apples are used for apple juice, some for cider, some for eating, and some for cooking. A cooking apple usually has the characteristic of being tart, and a sweet, juicy apple is great for eating, but not very good for making jelly or juice. So, different apple types have been combined, or bred, for different uses. A Catholic monk and scientist, Gregor Mendel, studied how these characteristics were passed from parent to child. If one parent apple tree produces eating apples with the characteristic of sweetness, and another parent apple produces eating apples with the characteristic of juiciness, it is possible to cross-breed, or combine, these two apples into a new tree that produces sweet, juicy apples. Characteristics from the parents are passed down, or **inherited,** by the new plant. We would call this new apple a **hybrid**, or a cross between two older types that makes a new type, or breed. Both animals and plants can be cross-bred to create a better breed.

In nature this happens also, only it sometimes looks random or accidental to us. The different kinds of food available, climate or weather, or diseases cause some animals and plants to produce more offspring, or offspring that is slightly different from others. This is why there are so many different varieties of plants and animals.

Graphs!

These are the most exciting things about science! Do you not believe me? Graphs make ideas clear to everyone who looks at them. Some graphs have even changed the course of science!

There are two very important things to remember about graphs. The first is that someone else must be able to understand it, and the second is that they should be beautiful. In this part of the course, we are going to learn about both of these.

Why do I say that graphs should be beautiful? Can you think of any reasons? What I think, and you can disagree if you want, is that graphs show the patterns of nature. When we consider that nature is the work of Our Lord, then anything that reflects nature should be beautiful. Further, we are the ones who make graphs, using our intelligence and creativity to understand the world

around us. We, of course, are also creatures of Our Lord, and our intelligence is one of the ways we are made in the image of God. So, when we use our intelligence wisely to make a graph, we are doing God's will by doing one of the things we were made to do. And, if you look at nature, you will see that everything in it is beautiful; even animals thought to be ugly are beautiful in their own way. So, to understand science and nature, we must strive to understand both truth *and* beauty in nature.

G. K. Chesterton said that if a thing is worth doing, it is worth doing badly. What he meant was that we should still try our best, even when we are just learning and we do not always do a perfect job. No matter! We try to do the best job we can, and in doing so, our work will be pleasing to God, even if it does not look great to us.

Words to Know

Having differences within a group or species is called **variability.**

A **characteristic** is a special difference that makes a plant or animal unique, so that it is not exactly like any other plant or animal in its species.

To **inherit** means to receive characteristics passed down by a parent to its offspring.

A **hybrid** is a cross between two older types of plants or animals that creates a new type or breed.

I will tell you something about trying to do your best work at any age. I drew some pictures when I was about ten years old, and then looked at them a few years later. When I was fourteen, I thought the pictures looked childish. When I grew up, however, I saw that they really were quite good (even the mistakes!) because I was trying to do my best. So, when you do a good job on your graphs, save the best ones in a scrap book.

Rules for graphs:

i. A graph should be easy to understand.

ii. Everything on the graph should be there for a reason.

iii. Everything on it must make sense.

iv. A graph should not confuse the reader.

Remember, the most important thing is for a graph to look beautiful. Then people will want to look at it to see what it says. A well-produced graph is just as much art as science!

The graphs on the following page are of birds at a feeder on January 6th. The graph on the left is a bar graph, and the graph on the right is a line graph.

Questions:
1. Do they both show the same data?
2. Which one is the best graph? Why?

Answers:

They both have the same data, but the one on the left is correct, and the one on the right is completely wrong. The line between the types of birds makes no sense. There are no half-birds between blue jays and chickadees as the line seems to indicate!

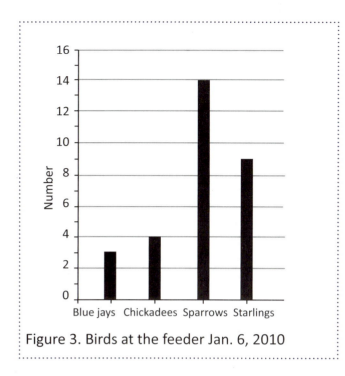
Figure 3. Birds at the feeder Jan. 6, 2010

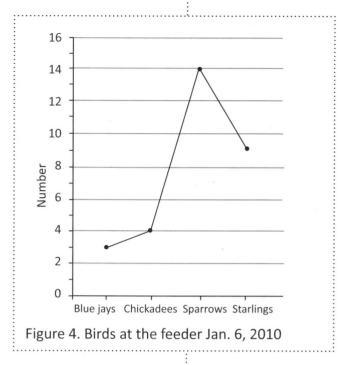
Figure 4. Birds at the feeder Jan. 6, 2010

Notice that with the graph on the left you can read all the information: there are nine starlings, three blue jays, fourteen sparrows, and four chickadees.

The bar graph shows the number of birds of different types, or categories. When data is organized in categories like this it is called categorical data, and a bar graph is used.

We can also use pie charts to record categorical data. Figure 5 is a pie chart that shows the same information about birds at the feeder. I do not think this type of chart shows the information quite as well, though.

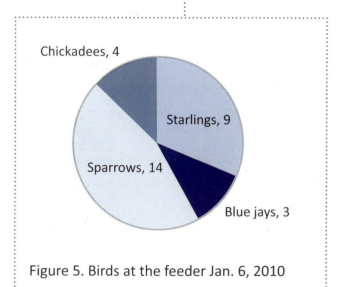
Figure 5. Birds at the feeder Jan. 6, 2010

Graphs Can Tell Lies

I am now going to show you how to tell lies with a graph! Rather, I am going to show you how some people try to fool their readers by using a graph that confuses them. You can look at graphs in the newspapers and magazines and see examples of this all the time. It is important to know these tricks so you will not be fooled yourself!

Figure 6 is a graph of the same data I collected from the bird feeder. In this graph, a bird is used instead of a bar to make it more interesting. But do you see what is wrong? In Figure 3, the original bar graph, the sparrow bar is exactly 3½ times the size of the chickadee bar.

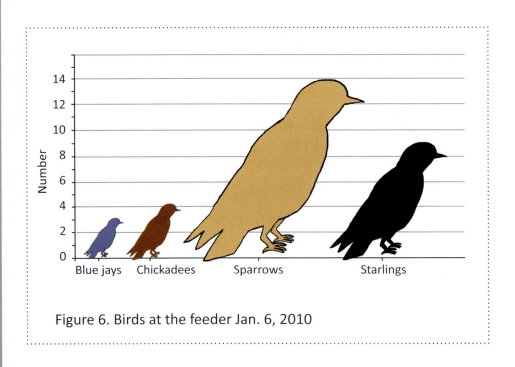

Figure 6. Birds at the feeder Jan. 6, 2010

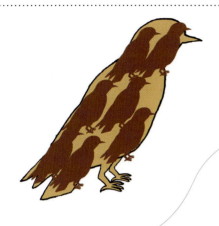

Figure 6 uses pictures of birds instead of actual bars. Do you notice how much bigger the sparrow bar is than the chickadee bar? The sparrow bar is actually 7 times bigger because it is wider! (Count the small birds. I think 7 chickadees fit into the sparrow bar.)

The trouble is that the graph should be based on height only, and the bars should be the same width. You should always check graphs to see that they have not made this blunder. Lots of people make this type of mistake, all too frequently!

Whether we are studying how beetles care for their young, or metamorphosis, or how creatures help each other by facilitation, learning to observe, keeping good records, and graphing our findings will all help us better understand the world around us. Are you ready to set up a bird-feeding station and start recording your own observations?

Let's put what we've learned to work.

Roll up Your Sleeves!

Lab Activity #1

Set up a bird-feeding station. You may use a bird feeder, or your station can be as simple as putting seed on a porch rail, a board, or even a plate on the ground. A windowsill also makes a good feeding station, for it will give you a close-up view of the birds. It may take a few days before the birds find your feeder and feel comfortable enough to come to it.

Purchase bird seed that is labeled for feeding many bird types and contains different types of seeds. Observe food preferences that may vary by bird type.

Using a notebook, make a note of the different kinds of birds and how many of each kind that come to the feeder. Because the birds will come and go, don't try to count them all day, or worry if one flies away while you are counting. Instead, count the birds in the morning and again around lunch time.

Keep a record for at least a week, and then create a graph or chart that shows how many of each kind of bird came to your feeder.

> Please remember to read all of the activities, right to the end of the chapter, even if you are assigned only one or two activities. Information contained in the activities is also instructional, and part of your lessons!

Behold and See 5 Student Workbook:

Worksheets for Chapter 2 begin on page 10.

Lab Activity #2: Flower examination and dissection

You remember that one type of facilitation is bees drinking nectar from flowers and then depositing pollen to pollinate the flowers. Pollination is what makes fruit and seeds form from the flowers. Without pollination, there would be very little food available for us to eat!

Let's study this diagram of a "generic" flower. Each type of flower is a little different from other flowers, so that a lily doesn't look exactly the same as a tulip, for example. But all the main parts are the same.

Starting at the top of the flower, the anthers drop **pollen** grains on the **stigma**, and then bees carry more pollen to the flower from other flowers. The pollen will make this flower fertile so it produces fruit.

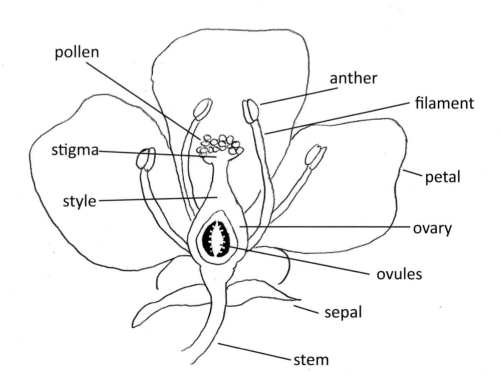

Words to Know

Pollen is the flower part that makes flowers fertile, or able to produce fruit.

The sticky part of the flower that catches pollen is called the **stigma**.

Words to Know

A **style** is a flower part that supports the stigma.

An **ovary** is the part of the flower that will become fruit.

A plant's **ovules** will become seeds.

Anthers are flower parts that produce pollen.

A **filament** is the flower part that holds up the anther.

Sepals are the parts of the flower that protect the bud.

The pollen grains sit on top of the stigma, which is sticky so the pollen grains don't fall off. If the pollen didn't stick to the stigma, the flower wouldn't be fertile, or able to bear fruit. That would mean no fruit for us to eat! Making the stigma sticky so the pollen stays on was pretty clever, don't you think? I wonder Who thought of that?

The **style** holds up the stigma; together, the stigma, style, and ovary are called the *pistil*.

The **ovary** will grow into fruit, and the **ovules** will become seeds inside the fruit. The seeds will start new plants.

The **anther** produces pollen; the **filament** holds up the anther. Together, the anther and filament are called the *stamen*. (Notice that the pollen-covered anthers lean somewhat over the stigma, so that some of the pollen can fall onto the stigma. This design can also help pollinate the flower.)

On the outside of the flower are the colorful *petals*, whose bright colors attract insects to the flower and help protect the inner parts of the flower. Also outside are the **sepals** that protect the flower bud before it opens, and the *stem* which supports the flower.

Select two flowers to examine and possibly dissect. Lilies and tulips and other large flowers are good choices for examination.

Examine your flowers. Can you identify the parts? How are the flowers the same? How are they different?

DO NOT DO THIS NEXT STEP UNLESS YOU HAVE YOUR PARENT'S PERMISSION AND HAVE AN ADULT HELPING YOU.

Place the flower on a firm, stable surface like a cutting board, and carefully cut the flower down the center. Can you identify the inner parts of the flower? Now do the same with the second flower and compare.

Let's Explore!: Expanding the Lesson

Pick a topic below to research online or in an encyclopedia. Draw a picture, write a paragraph, or tell someone what you learned.

- parts of a flower and their function
- pollination
- metamorphosis

Chapter 3

Food Webs, Resistance to Disease, and Conservation of Energy

God's majesty speaks to us by
the works of His almighty hands.
— Fr. R. H. Benson

Food Webs, Resistance to Disease, and Conservation of Energy

Food Webs

Your family probably buys its food from a grocery store, but have you ever wondered where plants and animals get their food? In the wild, plants and animals don't get their food from grocery stores, but through a system of **food webs**.

Food webs operate in a certain order, like this: First, at the top, is an animal called a top predator. The top predator is usually the biggest, or at least the strongest, animal within a habitat. The top predator is the one that eats other animals below it in the food web. Animals that eat other animals are also called **carnivores**, which means "meat eaters." Animals that eat only plants are called **herbivores**. And animals that eat both meat and plants are called **omnivores**.

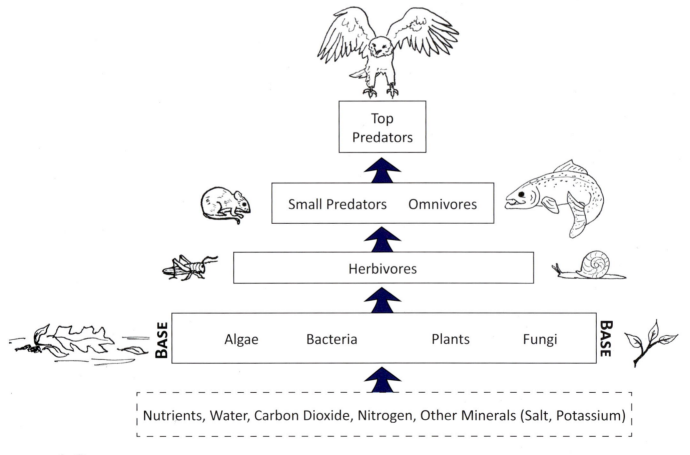

We human beings are top predators. That is one of our roles in nature, and in a food web.

In an ocean habitat, top predators are sharks, killer whales, sea lions, and large fish, such as the biggest cod. Top predators are sometimes eaten by other top predators. Sharks sometimes eat sea lions or sea otters, or large predatory fish. (Aren't you glad that you don't live in the ocean?)

In a pond habitat, the top predator can be a turtle, water snake, blue heron (which eats frogs and fish), or even an insect like a water beetle or dragonfly nymph (young dragonfly). Each different type of habitat can have a number of food webs in it.

On land, a top predator can be a wolf, coyote, weasel, bear, fox, or even a skunk or raccoon. Insect top predators can be ground beetles, tiger beetles, or adult dragonflies.

Top predators eat smaller predators, or herbivores, or omnivores. These, in turn eat smaller animals, until we get to the **base**, or bottom of the food web, which is made of plants, algae, bacteria, and fungi. The green plants at the base of the web use sunlight to make food out of non-living nutrients like water, carbon dioxide, nitrogen, and other minerals. Sunlight is the energy used by everything in the web so it can live. The energy of almost all living things (except things at the bottom of the oceans) comes from sunlight.

Words to Know

A **food web** is a system in which sunlight and both tiny and large plants and animals work together so everything in the cycle is nourished and fed.

Carnivores are creatures that eat only meat.

Herbivores are creatures that eat only plants.

Omnivores are creatures that eat both meat and plants.

The **base**, or bottom, of a food web is where the tiniest plants produce food from sunlight.

Words to Know

A **trophic level** is a level in a food web.

Primary producers are the creatures near the bottom of a food web.

Each level in a food web is called a **trophic level** and the further up an animal is on the food web, the higher nutritional value it has. Near the bottom are the **primary producers** or plants, fungi, algae, and bacteria, which live on non-living nutrients and water. Plants convert these nutrients into cellulose, which is the most common substance in plants. Cellulose makes the structure of plants, as we see in straw, wood, or leaves.

Human beings cannot live on grass or wood because we could not eat enough of it before we got full. But herbivores can and do eat these things. Since the nutritional value is not as high, though, these plant-eating animals need to eat lots of plant material to survive.

Predators, on the other hand, eat herbivores and other predators, so their food has much more protein in it. Because their food is of a higher quality, they don't need to eat as much of it to stay alive.

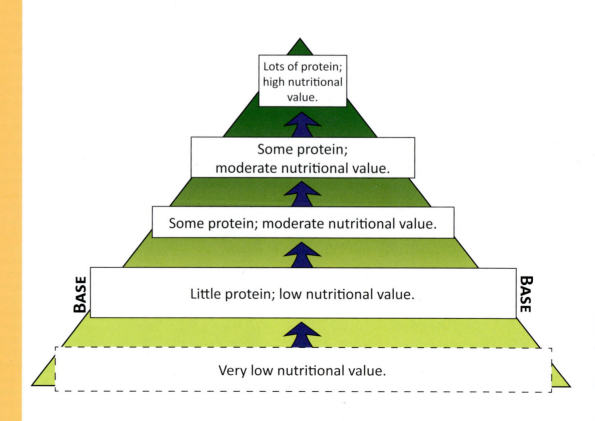

For example, I have dissected beavers and otters. The beavers are herbivores, and the otters are top predators. These two are about the same length, but beavers have a massive stomach the size of a soccer ball because they eat bark, a low-quality food. Otters are sleek and lean. They eat fish, a high-quality food, and have to be fast to catch them. Because their diets consist mainly of high-quality food, their stomachs are very small, smaller than a baseball.

On the other hand, the otter's heart is large. The otter needs a larger heart so it can get more blood to its muscles to chase fish and small rodents. An otter's heart is as big as my fist, and bigger than the otter's stomach. However, the beaver's heart is quite small, only a little bigger than a golf ball. The beaver's heart does not need to be as large as the otter's because it doesn't need as much blood pumped to its muscles. It is not hard to catch a tree!

This is the way food webs work in nature, in all the lakes, rivers, ponds, ditches, woods, fields, oceans, and swamps. And when animals and plants die, their bodies go back to the Earth to add nutrients to the soil and become food for more plants and animals. Can you see this in the diagram below?

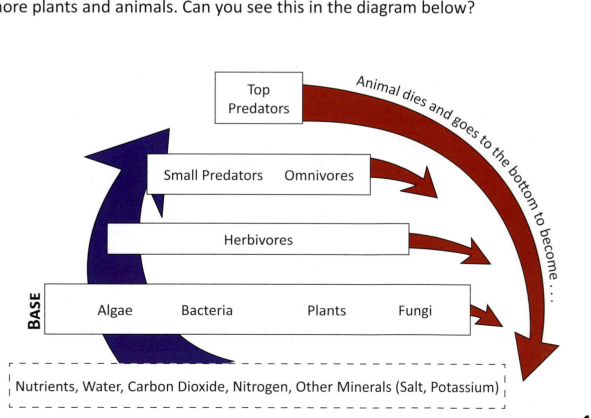

Words to Know

To **decompose** means to rot or break down into nutrients.

An **opinion** is how we feel about something, but it is not a fact.

Do you see that the food web is actually a never-ending circle, or cycle? The circle, or cycle, begins with the tiny nutrients at the bottom, that feed the algae, bacteria, and plants in the next level, which are eaten by the herbivores, which are eaten by the small predators and omnivores, which are eaten by the top predators. The last step in this cycle is death. The animals and plants **decompose**, or break down into the nutrients that feed the plants and begin the cycle all over again. We see this cycle of life in all lakes, woods, rivers, fields, and oceans.

So a food web is a food cycle that includes sunlight, tiny plants and animals, and larger plants and animals all working together so everything in the cycle is nourished and fed. Do you see how wisely God provided for His creatures? Unlike imperfect scientists, who can and do make mistakes, our perfect Creator never makes a mistake. Nothing is wasted or left out in God's plans. He has thought of everything.

Scientists Can Make Mistakes!

A swamp is a very important part of a food web, but here is a curious thing: scientists now call swamps wetlands. This is because some scientists think that the word swamp means a dirty place. Scientists thought that by changing the word to wetland, the swamp would seem more important. It is already important, even if we call it a swamp! If you look at the diagrams on the previous pages, you can see why. The swamp is where water gets cleaned by plants—the very low quality food—while they are changing nutrients into material for the herbivores.

Some scientists change words because of their **opinions**. It is important to always remember that scientists are people, and each one is just as different as anyone else is, so they have opinions, too. However, good science is not based on opinion, but on observation and recording of facts.

Some scientists like to hunt and fish (as I do), some do not; some live on farms (as I do), some in towns; some like old books and mystery stories (as I do), some do not; some go to church on Sunday—in fact, I know lots of scientists like myself who do—and others do not.

If you hear that a scientist has said something, make sure it is something about science before you believe his or her opinion. If the scientist is talking about something other than the actual facts of his scientific study, he may be telling you his opinion, rather than a fact. (A fact is something that simply is. If you see a mystery book on the table, it is a fact that it is a mystery book. But if you don't like mysteries, you might say that it's a boring book. That is your opinion. Your friend, who likes mysteries, might say that it's an excellent book. That, too, is an opinion. An opinion is how we feel or think about something, but it may or may not be accurate. An opinion is not a fact.)

Scientists have opinions, just like everyone else. Sometimes they make mistakes even when they should know better, especially if they are led by their opinions rather than facts. For example, do you know why there are gypsy moths in North America? An entomologist about 100 years ago in New England thought it would be a good idea to let some gypsy moths go and did not think it would do any harm. This was a BIG mistake. The caterpillars of the gypsy moths eat oak leaves and, as the gypsy moths and their caterpillars have spread and spread, eating as they go, they have cost millions of dollars to the timber industry and to individuals in lost trees.

Words to Know

Bioaccumulation is the increasing concentration of something (like chemicals) in animal populations.

Immunity in Insect Populations and Making Wise Choices

In our own country, we used to use a pesticide, or a chemical that can be used to kill insect pests, called DDT. It worked very well and killed lots of different insects. The trouble is that it began to make it so baby birds couldn't hatch, especially the offspring of eagles, hawks, and other birds of prey. DDT didn't hurt the birds right away when it was sprayed. Instead, insects and plants picked up the pesticide, and the birds and rodents and other herbivores ate the insects or plant seeds and the DDT in them. Eagles and hawks ate the rodents and smaller birds that had DDT in their systems. Then when the birds laid their eggs, DDT made the eggs' shells fragile so that the eggs kept breaking before their chicks were old enough to hatch. That is how the DDT accidentally killed baby birds.

With birds eating the insects that had been sprayed with DDT, and mice eating grains that had been sprayed with DDT, and then predators eating these mice and birds, the amount of DDT increased at each stage until it was very high in the top predators. This is called **bioaccumulation**. Bioaccumulation, or the increasing concentration of something (like chemicals) in animal populations, is why it is safe to eat a small fish from a polluted lake, but not an older big fish. Older fish have had many more years to pick up harmful chemicals in their bodies than younger fish.

So DDT got passed up the food web. Then in 1972, DDT was banned—outlawed, so no one could buy or use it—and not just in North America, but all over the world. The top predator birds were saved, and now we have lots of eagles and hawks in the United States and Canada, which is a very good thing.

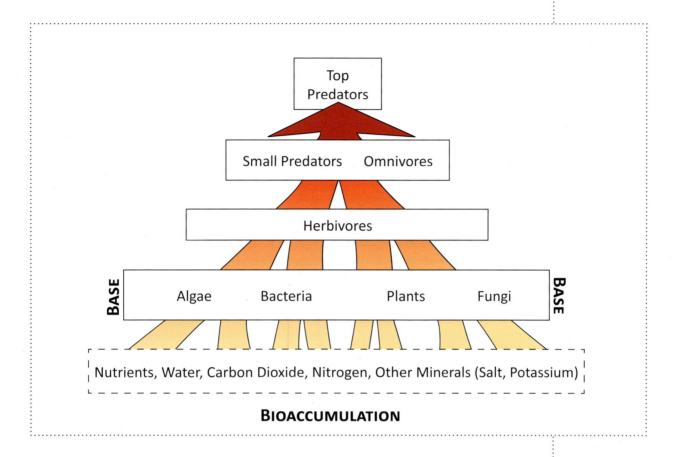

BIOACCUMULATION

But there is more to this story. DDT is actually one of the *safest* pesticides ever invented. Farmers could get it on their hands and it wouldn't hurt them. In World War II, the Allied forces, mostly U.S. soldiers, captured Naples, Italy. Because of the war, lots of the pipes for running water were broken and people could not bathe. Because of this, human lice (small parasitic insects that live in the hair on your head or on your body and make you itch) began to spread throughout the whole city. The lice carried a deadly disease called typhus. So the military set up washing stations and immersed everyone in DDT to kill the lice. It worked! Thousands of lives were saved throughout Naples, and a deadly typhus epidemic was stopped before it could spread. This is a *good* reason to use DDT!

DDT is very good at killing insects, and it is more or less safe for human beings and other animals so long as they don't eat it.

Words to Know

To be **immune** means you are able to "fight off" harmful substances or germs.

Immunizations are "shots" that keep people and animals safe from certain diseases.

Immunity is the ability of the body to "fight off" harmful substances or germs.

The problem with DDT is that it lasts a long time, which makes bioaccumulation possible. DDT also gets into the food webs by being washed into the ditches and eventually gets into the lakes and oceans. Some goes into the atmosphere, evaporating or being carried on small dust particles (called aerosols). The winds carry these, and drop them with the snow onto the Arctic region. Then it gets into the food web in the arctic habitat, and the seals, polar bears, and whales eat and drink it, and all the DDT gets stored in their fat and blubber.

Whale blubber now has so much DDT that the people who live in the Arctic and who used to hunt whales for food can no longer eat whale meat. As you can imagine, a whale could feed a whole village for a long time, and in the high Arctic there is not much else to eat other than whales, seals, caribous, and smaller game. So, whenever DDT is used, some of it eventually gets to the Arctic and makes it hard for the people who live there. This is a good reason to *stop* using DDT.

But consider this point. (There is a lot to learn about this! It is very complicated.) DDT was used so much to kill insect pests that many pests are now **immune** to it. To be immune to something means that you can't catch it, or it doesn't bother you very much. If you have immunity to a disease or to another substance, although it is still harmful for those who don't have immunity, it is not harmful to you. That is why scientists have developed **immunizations**, or "shots" that make people and animals immune to and therefore safe from certain diseases, especially diseases that are deadly.

For example, it is important that dogs get rabies "shots," as rabies can kill not only the dog, but also people who come in contact with it. If they have not been immunized against rabies, the people would die, too. But with a rabies "shot," they are immune to rabies.

Sometimes people and animals can have a natural immunity to some diseases. They may get sick, but not die, from a disease that would kill others. In the 1940's almost all house flies (Order Diptera: *Musca domestica*) could be killed with DDT. When DDT was used, which ones survived? The few that had a natural **immunity** to DDT.

For example, let us say that 1,000,000 house flies were killed in a southern town (such as somewhere in Florida) when DDT was used. There were lots of horses back in the 1940's in our cities and towns, and where there are horses, there are flies.

There were a lot more flies in those days than there are now, and house flies can carry horrible diseases. This is a very good reason to kill house flies and keep their numbers down to safe levels! But if you have 1,000,000 house flies, and 99.999% are killed, that still leaves 10 flies alive.

A kill rate of 99.999% means that only 1 out of every 100,000 flies survives. There would be so few that nobody would notice them. But each house fly can lay about 100 eggs, so their populations, or the number of house flies, double every month or so.

Now, do you remember that offspring inherit characteristics of the parents, like the hybrid apples in an earlier chapter? If some flies survived being sprayed by DDT, they must have had natural immunity. Because offspring inherit traits or characteristics from their parents, the house flies that were immune to DDT will produce house-fly offspring that have DDT immunity, too. So, like their parents, these flies will not be harmed by DDT, either. Let us include this in a model to see what happens.

Rules for setting up a house fly model:

1) House fly populations double every month.

2) DDT is applied to a whole town, then reapplied when fly population returns to its previous level.

3) House flies inherit DDT immunity from their parents.

4) When house flies get crowded enough, birds and other insects eat house flies to keep the number around a stable amount. Let this be 1,000,000 in our model town.

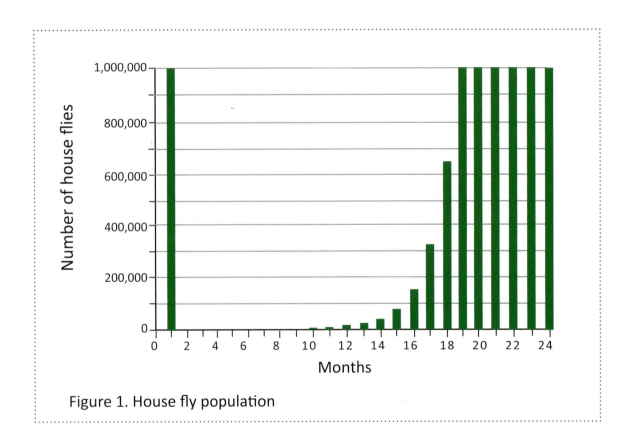

Figure 1. House fly population

Notice that the number of house flies is so small from month 2 to month 9 that you cannot even see them on the graph! When did the population get back to 1,000,000? That's right, after 19 months. (How many years is that?)

Table 1. Model of house fly populations in a town where the kill success is 99.999% so that only 10 survive.

months	house fly population	
1	1,000,000	DDT is sprayed
2	10 survive	
3	20	
4	40	
5	80	
6	160	
7	320	
8	640	
9	1,280	
10	2,560	
11	5,120	
12	10,240	
13	20,480	
14	40,960	
15	81,920	
16	163,840	
17	327,680	
18	655,360	
19	1,000,000	DDT is sprayed again, none are killed
20	1,000,000	DDT is sprayed again, none are killed
21	1,000,000	DDT is sprayed again, none are killed
22	1,000,000	DDT is sprayed again, none are killed
23	1,000,000	DDT is sprayed again, none are killed
24	1,000,000	DDT is sprayed again, none are killed

All of the house flies are now immune to DDT. This is exactly what happened, and it was first noticed in Florida in the late 1940's and early 1950's. DDT immunity spread throughout all of the United States and Canada, so that, as far as we know, all house flies are now immune to DDT! This means that if the fly population were spreading a disease and we wanted to kill the house flies, DDT would be useless and, unless we had a different pesticide to kill the flies, we would have a big problem on our hands.

The fact that pests can inherit immunity is one of the problems with any pesticide. (Notice that at least some of the house flies already had to have this immunity for this to happen.) Whenever pesticides are used, there is the risk of building immunity into the pests.

Can you think of a way to solve this problem?

Imagine you have an orchard, and you are losing apples to pests that damage your fruit. You have tried to spray, but the pests are becoming immune to the pesticide, and you know that you will soon be dealing with pests you cannot kill.

What if you kept some apple trees nearby that you did not spray, say along the fence. These would still have pests that would lay their eggs on your apples every year. Yes, they would do some damage to the trees and fruit if you did not spray them with a pesticide. But they would not build up an immunity to the pesticide, either. You would be raising your own pests so you would still be able to kill them when they attack your crops. They would not have inherited immunity because you do not spray them everywhere. This seems illogical to some people, but it works!

Here is the new rule: only spray insects that are pests.

To better understand this, let's make a model. Here are the rules for this new model:

1) Pests fly around every spring looking for mates and new apple trees to infest.

2) Females lay eggs on apple blossoms.

3) The farmer sprays the orchard which kills 99% of the pests; in the fall he can harvest a clean crop.

4) The farmer does not spray trees along the edge of the orchard.

5) Pests on the trees along the edge of the orchard double in number.

6) The next spring, pests from the unsprayed trees along the edge fly into the middle of the orchard and breed with any immune pests still alive so that immunity is not inherited.

7) Birds and dragonflies and other predators keep the number of pests in the orchard from becoming greater than 10,000.

We will start with 10,000 in Field A and Field B.

Table 2.

Year	Season	Field A pests on unsprayed trees	Field B pests on sprayed trees	
1	spring	10,000	10,000	
1	summer	10,000	100	Middle of orchard is sprayed, only 1% survive.
1	fall	10,000	100	Farmer can harvest a clean crop.
2	spring	5,000	5,100	Half the pests from unsprayed trees fly to sprayed trees.
2	summer	10,000	51	Pests double on unsprayed trees; middle of orchard is sprayed, only 1% survive.
2	fall	10,000	100	Farmer can harvest a clean crop.
3	spring	5,000	5,100	Half the pests from unsprayed trees fly to sprayed trees.
3	summer	10,000	51	Pests double on unsprayed trees; middle of orchard is sprayed, only 1% survive.
3	fall	10,000	100	Farmer can harvest a clean crop.
4	spring	5,000	5,100	Half the pests from unsprayed trees fly to sprayed trees.
4	summer	10,000	51	Pests double on unsprayed trees; middle of orchard is sprayed, only 1% survive.
4	fall	10,000	100	Farmer can harvest a clean crop.

The result is a good, clean crop each fall and no pesticide immunity. In this way, wise use of the spray produces a good crop, prevents pests from becoming immune, and is not wasteful and costly. Best of all, this method does not harm other animals or insects that can help control the pests. Does this make sense? The great thing is that you can protect your crops, and it costs less because you use less pesticide.

How does this apply to DDT? Here is how. Because so much DDT had been used by the 1970's that it was killing birds, most of the countries of the world agreed to stop using any DDT at all. That was fine for us who lived in North America, since the DDT had stopped working anyway so it was no big loss. But the southern countries where there were lots of mosquitoes that carried malaria also could not use DDT. People in these countries all over the world began to get sick again and die from malaria, just when malaria was almost under control everywhere.

See the problem? Too much DDT caused people who live in the far north to lose their food source because they couldn't hunt whales and seals, and it caused lots of our insect pests to become immune to DDT. But when DDT was banned, the people who lived in the tropics got sick with malaria. What is the right thing to do?

It's easy to figure out, if you use good logic and remember some basic rules that really go right back to creation. In the very first chapter of Genesis, God made man in His image. Nothing else in creation was made in His image: not rocks nor animals, neither plants nor insects. Man alone was made in the image of God. In addition, only man was given the gift of reasoning. And with this gift, God also gave mankind an important and solemn job: to subdue and have dominion over all the Earth. This means that God made man the most important of all His creatures, and that man has the responsibility of wisely caring for all else that God created. So man needs to think and look for God's guidance about how he uses the Earth and the creatures on it, so he can figure out the right thing to do.

Here are the rules for figuring out what to do:

1) **Animals and plants are important, but people are more important than animals and plants.**

2) Using too much DDT causes harm to people in the north.

3) Banning DDT causes harm to people in the south.

The solution? Use DDT wisely to save lives, just like the farmer used pesticides wisely on his orchard to save his fruit. Does this make sense? I think it does, and so do other biologists who live in both the north and south. Overreacting by banning DDT in the 1970's was just as big a mistake as using too much in the 1940's. As I said previously, there are lots of different scientists, and lots of different opinions and points of view, and not all of them are right! But we can use the reasoning ability God gave us to think of good solutions.

Now, look at the house fly graph again (Figure 1). People are a lot different from house flies, right? Certainly we are! However, many of the rules of nature apply to us as well as to insects, animals, plants, and everything else. In the next section, we explore how immunity works with people, too. First, answer this question: Based on Figure 1 and Table 1, what can you say about a house fly in month 19?

If you said that it was immune to DDT, you were right.

Resistance to Disease: Immunity in Human Populations

So, based on this reasoning, I am going to tell you something I know about you, your family, and all your friends. You and I come from a long line of survivors, some of the strongest people in all of history. We are all the descendants of people who survived hundreds, even thousands of disease epidemics that occurred in the past. (I sometimes find this kind of humorous, especially when I get a cold or poison ivy). Why did our ancestors survive, when so many others died? It was because the survivors had natural immunity, or stronger immune systems.

In the Middle Ages, a terrible plague called the Black Death swept through Europe. The Black Death was spread by fleas (parasites!) that lived on rats that were infected with the plague, and the fleas would bite people and infect them with the plague. Almost half of the people living in Europe and Asia died from this terrible plague, and it kept coming back every couple of generations. These plagues were horrible, but every time the plague came back it became less severe, not because the disease changed, but because the immunity of the population changed. People whose immune systems couldn't fight the plague died, so they didn't have children. People with stronger immune systems survived and had children who were naturally immune to the disease.

By the time the last plague went through Europe, most people who got sick recovered instead of dying. Now, although the plague still exists in some rats in wilderness areas of California and Mexico, very few people ever get sick from it, and those who do can go to the hospital to get well again. Do you see what happened? Those who survived had children—offspring—who inherited immunity from their parents. Their bodies "recognized" the plague germs because of the immunity, so they could fight off the germs.

Everyone alive today whose family came from Europe or Asia is a descendant of the people who survived the plague. All during

Fresh Water!

In most of North America, Europe, Australia, and other modern countries, we now have faucets with fresh clean running water in our modern houses. Yet in a way, we are living like the early humans did with clean running water nearby, just as if we had a small stream running beside our camp! Each time I turn on a cold water tap, I think of my long-ago ancestors who would have lived in a one-room house made of rocks and sticks covered in mud (this is called wattle and daub) and with a roof of animal skins or straw thatch. They had a fire in the middle of the floor, and they would have drunk clean, running water from a stream beside their house.

Do you think that we are so very different from people who lived long ago? My house has wooden stud walls covered in plaster. Plaster is just a kind of dried mud made from a rock called gypsum with water added to it. Lumber is just a large straight stick of wood. Do you have steel in your house? Steel is just a rock that has been melted and cooled down again. The fire that heats my house is divided into two rooms. One of the fires we call an oven, and the other a furnace. Really, our houses are still made of sticks covered in mud with a fire inside. And, like the water our early ancestors drank, the running water is once again fresh and clean. But with the help of science, we now understand why we need to keep the water clean, and how to do it!

these years other diseases such as typhoid fever, diphtheria, cholera, tuberculosis, polio, influenza, and lots of others with no names were killing people in Africa, Asia, Europe, and North and South America. We all come from a long line of ancestors who survived these diseases, too.

It is not that our ancestors were better people. It is just that some people's immune systems were able to fight these diseases and other people's immune systems could not. Those with the toughest immune systems survived, while others died. Everyone alive in the world today is a descendant of those people who have the toughest-of-the-tough immune systems.

Part of the reason that we are healthy now is because we have clean water to drink and soap to kill germs and keep us clean, and screens on our windows to keep the flies out, and modern medicine. But much of the reason for why we are healthier now is because we have inherited strong immune systems.

Why did the people in the Middle Ages get so sick? Our earliest ancestors hunted and gathered plants for food. In those older times a family group would live beside a stream filled with clean running water. Then, every so often, the group would move on following the game through the different seasons. Because they didn't stay in one place for very long, the water sources did not become contaminated.

Then came the agricultural revolution, when human beings first began to farm. Animals and people stayed in the same place, so it was no longer safe to drink from the local streams. Even well water got polluted sometimes. Long ago, people didn't have bathrooms like we have now, and human waste spread into the streams and rivers that people drank from. Farms are a very good thing, because they can feed more of our hungry brothers and sisters across the world. And most modern farms operate in ways that keep the water clean. But many of the diseases of history started with water that was polluted not by pesticides or industry, but by animal or human waste.

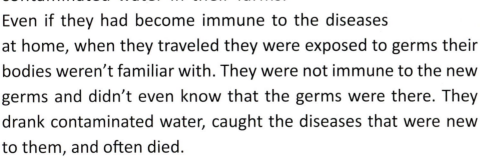

Eventually people started travelling again, this time to trade with other countries, go on Crusade, or find new land to farm. But this only made things worse, because they spread the new diseases that had been started by the contaminated water in their farms. Even if they had become immune to the diseases at home, when they traveled they were exposed to germs their bodies weren't familiar with. They were not immune to the new germs and didn't even know that the germs were there. They drank contaminated water, caught the diseases that were new to them, and often died.

Words to Know

An **atom** is the basic building block of the universe, and is the smallest "whole piece" that a thing can be broken down into.

Molecules are very tiny substances that are made up of one or more kinds of atoms.

At the same time, people already living in the areas that the Crusaders or settlers were coming into had no immunity to the germs that the Crusaders or settlers brought with them. As they traveled, the new settlers contaminated the local water supply with the germs they brought with them. Of course, people didn't do this deliberately, but because of their ignorance. Still, people already living in these lands caught the "new" diseases. Diseases spread from one group of people to another. Some died, but many survived these new diseases, and built up immunity to them. Little by little, people built up immunity and scientists learned how to treat the terrible diseases that used to kill lots of people. This is a good reason to keep studying science!

In many parts of the world, however, people still don't have clean water coming from their faucets. They still drink from rivers, streams, and wells where the water is not pure. Sadly, the lack of clean water still causes many deaths from disease in poorer countries of the world. The good news is that many Catholic organizations are working in these countries to dig wells with clean water. Maybe you would like to help dig wells for the poor in these countries when you grow up.

Conservation of Energy and Materials

You have learned how creatures help each other through facilitation so that their lives are easier, and you have learned how animals rely, or depend, on other animals for their food. Can you see how this works in the food webs that we studied earlier in the chapter? God has carefully planned His world to be one in which we all depend on other parts of His creation. Nothing is wasted in His plan. This is also true of the basic building blocks—carbon, hydrogen, and oxygen—that God used in creation.

Do you know what wood is made of? It is made of carbon dioxide and water put together using sunlight! Carbon dioxide is a gas. It is what you breathe out, and it is represented as CO_2, meaning one carbon and two oxygen atoms bonded together. You probably know that water is represented as H_2O, meaning two atoms of hydrogen and one oxygen atom bonded together. (An **atom** is the smallest "whole piece" that a thing can be broken down into. Atoms can be put together to make slightly larger building blocks called **molecules**. Even though they are very tiny, molecules are not the smallest thing, because molecules can be broken back down into atoms.)

In a way, wood is stored sunlight energy. When wood rots, and fungus and mold grow on it, the sunlight energy is slowly released. Some of the sunlight energy is turned into fungus and mold, and the rest is used to break the wood back down into H_2O (water) and CO_2 (carbon dioxide).

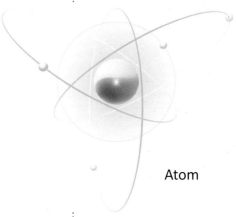

Atom

When wood burns, the fire is just the sunlight energy being released quickly, and the wood is turned back into H_2O (water, which is what most of the smoke is) and CO_2 along with a bit of ash and charcoal. Charcoal, of course, is just wood that couldn't finish burning before the fire went out.

Remember the food cycle, and how the nutrients fed the plants that fed the animals, which died and became nutrients

Words to Know

Conservation of energy is the scientific law that says energy can be neither created nor destroyed.

to start the cycle all over again? In that food web, we saw that nothing is wasted in God's plan, that He created everything, and that all He created is part of His grand design and system.

There is a "law" that you will learn more about later in physics, which is another branch of science. That "law" states that energy can neither be created nor destroyed. This is known as the **conservation of energy**. Energy can change form—sunlight energy can become plant energy which can become energy for animals—but the energy is not destroyed. This is an important part of God's plan to provide for all His creation.

God gave man an intellect so he can think about these things and understand better how God's creation works. Let's put our intellects to work, and think about some questions related to conservation of energy and materials.

1. If I have a pail of sand, and then dump it out to make a sand castle, will the amount of sand be more, the same, or less than there was in the pail?

2. If I bring a pail of sand to a sandbox and use it to make a sand castle, and then fill a second pail exactly the same size with different sand from the sandbox, will I have more, less, or the same amount of sand in my new pail?

3. If I order the sand from Alaska to fill the pail, will that change the amount of sand in a full pail?

4. If I fill the pail slowly, will that change the amount of sand in a full pail?

5. If I fill the pail quickly, will that change the amount of sand in a full pail?

These seem like simple questions, and they are. But let us try this same set of logic questions, but instead of a pail we will use a tree, and instead of sand, we will use CO_2 (carbon dioxide). We will assume that our tree grows to the same height and diameter (width of the trunk) every time.

1. A tree grows, using CO_2 to make the wood, bark, and branches. Then I cut it down to light a fire in a wood stove or fireplace. Will the amount of CO_2 be more, the same, or less than there was in the tree?

2. If a full grown tree is cut down and used for firewood, and another tree exactly the same grows where the first tree was, will the amount of CO_2 in the new tree be more, the same, or less than there was in the old tree?

3. If the new CO_2 comes from Alaska, or Holland, or Nova Scotia, or California, will that change the amount of CO_2 in the new tree?

4. If I do not use the tree for firewood, but instead it just falls down and rots slowly and a new tree grows, will the amount of CO_2 be any different in the new tree?

5. Does rotting slowly or burning quickly change the amount of CO_2 in a new tree?

6. What if it takes ten years for a new tree to start growing. Will that matter? If it takes 100 years? 1000 years?

7. Now here is a tricky one. What if the tree was buried for 100,000 years, and then dug up and burned, and a new tree grows, will that change the amount of CO_2 I have?

This is really important! When anything grows, it uses the material that is all around it, and when it dies, that same material goes back to be used by something else. You have seen this cycle in food webs, and also now with conservation of energy. These are both part of God's good plan for supplying the needs of, and caring for, all that He created.

Man, too, can care for God's creation by wisely using the materials that God provided for our use. God gave mankind the gift of reasoning so that we can think of ways to protect crops and animals and people from disease, ways that care for all of God's creation.

Let's put what we've learned to work.

Roll up Your Sleeves!

Please remember to read all of the activities, right to the end of the chapter, even if you are assigned only one or two activities. Information contained in the activities is also instructional, and part of your lessons!

Activity #1

Many zoologists, including myself, prefer to study live animals rather than dead ones, and sometimes the only way is to keep them alive for observation is in a lab. I have two labs, one in my living room, and one in the garage. A lab is just the place where you can do your studies, and it does not need to be complicated.

In my lab I have a magnifying glass, an old microscope, some boxes with pinned insects in them, and a bookshelf for my books on how to identify insects, weeds, trees, rocks, and other things like that. In the summer, my lab has insects in cages, and sometimes we keep a turtle. We put him in a washtub in the kitchen at night, and in a wading pool during the day on the back porch.

Now that you understand food webs and how much you can learn about animals by observing them, this would be a good time to research and observe some animals on your own. Are you ready?

You will need your notebook to record how you caught your study animals, and you will need a pencil so you can make a drawing of your specimen. And, most important, you need to watch them to see what they eat.

Behold and See 5 Student Workbook: Worksheets for Chapter 3 begin on page 19.

65

Words to Know

When an insect **molts**, it sheds its skin.

What kind of animal do you want to research?

If you decide to capture your own specimens, you may wish to begin with something that you can see feeding, so you will have an idea what to feed your insect or animal. For example, if you want to research butterflies, you might catch them drinking flower nectar. Then you will know which flowers to keep in their container so they will have food.

If you aren't sure what your animal eats, look in an encyclopedia or online. Then offer your animal different kinds of food so you can see what they reject and what they like best.

Even though they are slimy, snails and slugs are easy to catch and keep. You can find them under rocks and logs and in your garden or flower bed. In fact, they are probably eating your flowers or vegetables, so you will already know what they like to eat!

If for some reason you can't catch your own animals, you can find mealworms and crickets in pet stores. They are inexpensive and interesting to observe.

Mealworms

Mealworms are interesting creatures to study. They are beetle larvae, so you can observe them pupate and turn into adult beetles, usually in a month or two if they are kept in a warm place. Make sure that you keep them in a container that has holes in the lid.

Mealworms can eat bread crumbs or oatmeal, as well as vegetable or fruit peels or pieces. (Mealworms take their moisture from the vegetable and fruit pieces.)

Put bran or dry oatmeal, or even some leaves, into the

mealworms' container so they will have a place to hide. Like the beetle grubs mentioned in an earlier chapter, your mealworms will **molt**, or shed their skins as they grow. When you see the shed skins, make note of when and how often they molted before pupating.

Be sure to draw a picture of your mealworms at various stages in their development. Do they have legs? How many? How long are the mealworms? Do they stay the same length or size?

Crickets

Crickets are also fun to raise. It is a good idea to get three or four crickets, so you will have at least one male and one female. (Adult female crickets have a long "tail" or stinger-like tube called an ovipositor, which is used to lay eggs in the ground. Males don't lay eggs, so they don't have an ovipositor.) If you put a couple of inches of dirt in your container,

it is possible that your female crickets will lay eggs in the container.

Crickets like many different kinds of foods. Offer them a variety of grain-based foods, fruits and vegetables, and even small pieces of meat. Keep track of which foods they like best, and which they reject.

Crickets also molt, or shed their skins as they grow. Record when they molt, and how often.

Draw your crickets. How many legs do they have? What other body parts do they have? Use the encyclopedia or online research to find out the names of the body parts that you see, and learn more about the habits of your research cricket.

Activity #2

Here is a simple habitat. Can you guess the rest of the food web?

habitat	**primary producer**	**herbivore**	**predators**
farms	_____	_____	people

4
Chapter

Physiology and Introduction to Biochemistry

The very order, disposition, beauty, change, and motion of the world and of all visible things proclaim that it could only have been made by God . . .

— St. Augustine

Physiology and Introduction to Biochemistry

Words to Know

Biochemistry is the study of chemical composition and reactions in living things.

Chlorophyll is a pigment that causes the green color in plants. Chlorophyll is necessary for photosynthesis.

Photosynthesis is the process by which plants use sunlight to turn carbon dioxide and water into sugar with the help of chlorophyll.

Photosynthesis

You remember that sunlight is a source of energy. Did you know that plants use sunlight to make sugar, and can link the sugars together into starch or cellulose? The plants achieve this process with the help of a pigment called **chlorophyll**, which is what makes plants green. This pigment traps the sunlight so that plants can use it to create energy in the form of a simple sugar.

The process of using sunlight to make sugars is called **photosynthesis**. The "photo-" part of the word means "light-" and "-synthesis" means "-construct" or "-assemble."

So, photosynthesis just means to assemble using light. The light is sunlight, and the raw materials are carbon dioxide and water. In the process, the plants also release oxygen into the air. This is a good thing, because human beings and animals both breathe out carbon dioxide and breathe in oxygen. If it weren't for the plants taking in carbon dioxide and releasing oxygen, we would eventually use up all the oxygen and fill the air with carbon dioxide. But this cannot

happen, because the plants keep releasing oxygen through the process of photosynthesis. And it all goes back to God's brilliant plan for sunlight! Do you see that, in God's design, nothing is truly wasted, but everything has a purpose?

How is this sugar from plants, called **glucose,** made?

Glucose is made up of carbon, hydrogen, and oxygen atoms, in these amounts: $C_6H_{12}O_6$. The "C_6" stands for six carbon atoms, the "H_{12}" stands for twelve hydrogen atoms, and the "O_6" stands for six oxygen atoms.

When carbon, hydrogen, and oxygen atoms are put together, they form a molecule of the sugar glucose. Sugar is a kind of **carbohydrate** that is produced by plants. In fact, you probably had some carbohydrates for breakfast this morning in the form of cereal or toast!

Atoms, Elements, and Molecules

Atoms and molecules are the "building blocks" of the universe. If you broke a piece of iron into the tiniest piece of iron that could exist, you would have one atom of iron.

Atoms are so small that they can only be seen with the most powerful microscopes. Atoms are put together with other atoms to make molecules, another kind of "building block." Most molecules are compounds, which means that they are made of more than one kind of atom. A compound is a mixture of different kinds of atoms that makes something completely new. Water is a good example of this, because it is a combination of two hydrogen atoms and one oxygen atom. Hydrogen and oxygen are the elements that water is made of. An element is made up of only one kind of atom, so it cannot be broken down into simpler chemicals.

Oxygen, carbon, and hydrogen are three important elements. Water, carbon dioxide, and glucose (sugar) are important compounds.

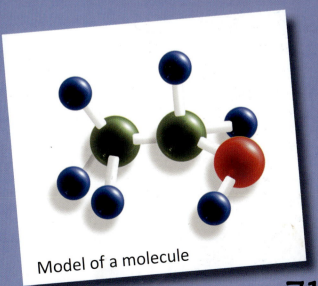

Model of a molecule

Words to Know

Glucose is a type of sugar, and is made up of carbon, hydrogen, and oxygen atoms.

Carbohydrates are types of sugar that are produced mostly by plants, and are made of carbon, hydrogen, and oxygen.

An **element** is a chemical or substance that cannot be broken down into simpler chemicals.

Bonds join atoms together to form molecules.

Let us look at this word: *carbo–* means carbon; *hyd–* means hydrogen; and *–ate* on the end of the word means there is oxygen. The important difference between types of carbohydrate molecules is the number of carbon atoms each has. Glucose has six carbon atoms bonded together in a kind of chain, so we call it a 6-carbon chain. It looks like this:

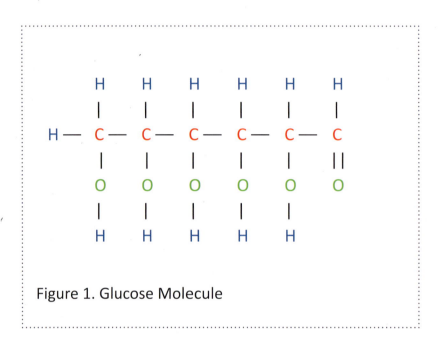

Figure 1. Glucose Molecule

Please draw Figure 1 in your notebook. It is important to see how molecules are put together, and it would be helpful to memorize this.

The letters stand for the individual atoms of each **element**. In the diagram, "C" stands for a single carbon atom, "O" for an oxygen atom, and "H" for a hydrogen atom. The lines are called **bonds**. Bonds join the atoms together to form molecules. (To picture this, think of the atoms as beads and the bonds as chain links that fasten one bead to another.) Can you see the "chain" of carbon atoms in the middle of the glucose molecule (Figure 1)? All the oxygen and hydrogen atoms are bonded to this chain.

Copy the tables below in your notebook.

Questions: How many bonds are attached to each carbon atom? (Count the lines coming from each one). How many bonds are on each oxygen atom? On each hydrogen atom?

Put these numbers in your table. You can keep filling in the table as we go along.

Element	Symbol	Number of bonds
carbon	C	4
hydrogen		
oxygen		

Molecule	Symbol	Number of each kind of atom
glucose	$C_6H_{12}O_6$	C: ___ H: ___ O: ___
carbon dioxide	_____	C: ___ O: ___
water	_____	H: ___ O: ___

Here are three other chemicals, their chemical symbols, and their diagrams:

Chemical	Chemical Symbol	Diagram
water	H_2O	H—O—H
carbon dioxide	CO_2	O=C=O
oxygen gas	O_2	O=O

(Oxygen gas is the stuff in the air that you and I breathe; you can see that it is made out of two oxygen atoms.)

Draw diagrams of oxygen, carbon dioxide, and water in your notebook just like the ones on the previous page. Use your diagrams to fill out the rest of the table of elements, symbols, and bonds.

Now look at the glucose diagram (Figure 1). Do you notice something curious? Glucose can be broken down into water, oxygen, and carbon dioxide.

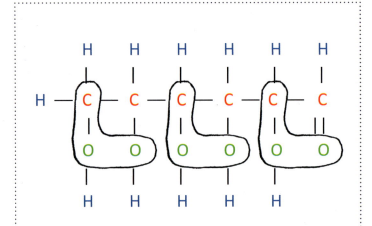

Figure 2. Glucose can be broken into carbon dioxide molecules.

How many carbon dioxide molecules can you get from glucose (Figure 2)? To answer this question, I circled groups of two oxygen atoms and one carbon atom on my diagram.

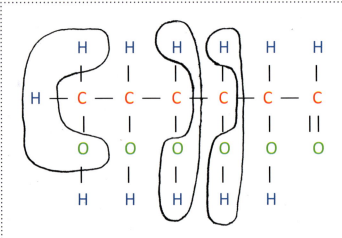

Figure 3. Glucose can be broken into water molecules.

How many water molecules can you get from glucose (Figure 3)? To answer this question, I circled groups of two hydrogen atoms and one oxygen atom on my diagram.

Try this on your own diagram to see if you can find more water molecules than I did (there should be six total).

Which two elements are left over?

You're right, carbon and hydrogen in Figure 2, and carbon in Figure 3.

Which element do we run out of in both figures?

That's right, oxygen.

To change all the carbon and hydrogen atoms in a single glucose molecule into water and carbon dioxide, what do we need more of?

Right again! Oxygen!

If I tell you that you eat glucose (sugar) for energy, and breathe in oxygen, and breathe out carbon dioxide and water, can you figure out what is happening in your body?

You are using the oxygen to break down the glucose into carbon dioxide and water.

How many more oxygen atoms are needed to turn all the carbon and hydrogen in the glucose molecule into carbon dioxide and water?

Here is my attempt at figuring this out. You try this too:

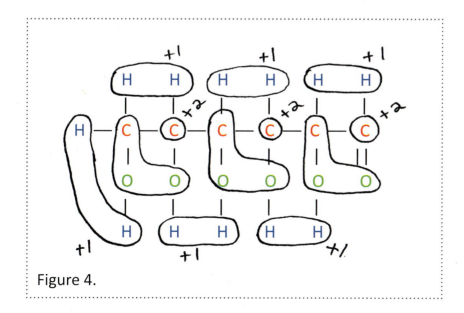

Figure 4.

Words to Know

Respiration is the process of using oxygen to release energy.

A **chemical equation** is the "recipe" for how chemicals are put together to form a material or cause a chemical action.

So how many more oxygen atoms were needed?

I count 12 more. That means six O_2 molecules are needed.

We can write this out as:

$$C_6H_{12}O_6 + 6\,O_2 \rightarrow 6\,H_2O + 6\,CO_2 + \text{energy}$$

glucose + oxygen → water + carbon dioxide + energy

The arrow means "produces."

How many carbons are on the left? How many on the right?

How many hydrogens on the left, and how many on the right?

How many oxygens on the left, and how many on the right?

What is left over on the right after the glucose is turned into water and carbon dioxide?

That's right, *energy!* Do you see what has happened? When you eat a piece of bread, you are eating thousands of glucose molecules, which plants have packed full of energy. When your body uses this glucose for food, it uses oxygen to break each glucose down into water, carbon dioxide, and energy. Then your body uses the energy to move and grow! This chemical process of getting energy is called **respiration**, meaning that oxygen is used to release energy.

Where do the plants get the energy they need to make glucose? From the sun! The energy is sunlight. So, all the energy in our body is sunlight that has been released into our cells! Almost every living thing uses energy from sunlight. (The exceptions are those things that live in the bottom of the sea or at the edge of volcanoes; they use heat from the Earth as their energy.)

Take another look at the **chemical equation**, or "recipe," for respiration:

$$C_6H_{12}O_6 + 6\,O_2 \rightarrow 6\,H_2O + 6\,CO_2 + \text{energy}$$

glucose + oxygen → water + carbon dioxide + energy

Now study the chemical equation for photosynthesis:

$$6\,CO_2 + 12\,H_2O + \text{sunlight} \rightarrow C_6H_{12}O_6 + 6\,O_2 + 6\,H_2O$$

carbon dioxide + water + sunlight → glucose + oxygen + water

Can you see on the left side of the equation how the plants use sunlight to combine carbon dioxide and water? The right side of the equation tells us that the result of photosynthesis is glucose, oxygen, and water. Since the plants cannot use the oxygen, they "breathe" it back out into the air, and then human beings and animals use it in the process of respiration. (Look on the left side of the respiration equation.)

In the same way, the carbon dioxide that we breathe out (look on the right side of the respiration equation) is just what plants need to continue the process of photosynthesis! This is one of the most important examples of facilitation.

Do you remember learning about the conservation of matter and energy in the last chapter? Count up the carbons, oxygens, and hydrogens to make sure there are the same number on each side of the photosynthesis equation.

I trust you have heard of a longbow? It was the bow used by Robin Hood and his friends such as the Franciscan, Friar Tuck. Longbows were also used at the famous battle of Agincourt. The history of longbows still affects us today: it is because of the laws governing the use of longbows that the United States and Canada have the right to bear arms. So now you have a little history and government mixed with your science. Isn't it interesting that, as we study and learn more and more, we can see connections across so many subjects?

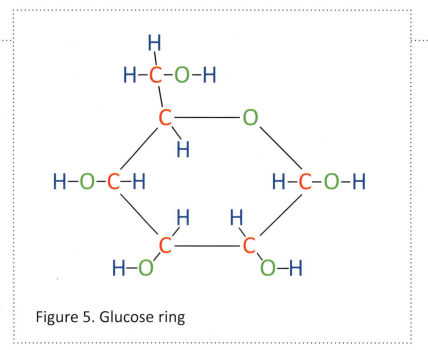

Figure 5. Glucose ring

Turning Glucose into Strong Building Blocks

Now, let's go back to glucose, the simple sugar that provides our energy. Glucose can be a chain, or it can turn back on itself and form a ring. Figure 5 is what it looks like when it is a ring.

Count the oxygens, carbons, and hydrogens to make sure there are the same number as there were when the glucose was a chain.

Can you guess what kind of material can be made with a ring instead of a chain?

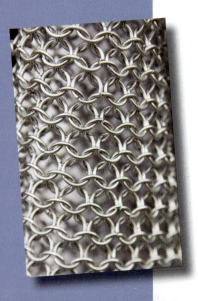

Here is a clue: Have you ever heard of chainmail? Chainmail is the armor that knights used to wear. It is made of lots of iron rings linked together into a kind of cloth made out of iron. Because of all the rings, it is very flexible and strong. In fact, it is so strong that crossbow arrows cannot go through chainmail. The only kind of bow that can pierce chainmail is the longbow.

Now let's think of glucose rings being linked together into something like chainmail. For chainmail, using smaller rings makes a finer chainmail which is more flexible, and still very strong! Rings of glucose are much, much smaller than any chainmail rings. They are so small, you can't even see them under a microscope! If rings of glucose were linked together the way chainmail is, the result would be very tough and flexible stuff!

In plants, a sheet of glucose "chainmail" is called **cellulose**, which is another name for wood fiber. This is what plants are made of, and it is what makes trees so strong. With a few changes, a different kind of glucose chainmail becomes **chitin**, which is in the "armored shell" of insects—tough stuff! And God thought of it long before chainmail was invented!

Cellulose rings are tightly linked together, and they are hard to break apart. This is why wood cellulose is hard to eat. We cannot eat it, even though it is made of sugar. However, termites, fungi, and bacteria can eat wood cellulose.

When trees grow, they use sunlight to make glucose out of water and carbon dioxide. You know how this works now. But what happens when you burn paper or wood? Heat and light are created. The heat and light produced are actually just the sunlight that made the plant grow being released very quickly.

The Carbon Cycle: Many Types of Fuel Begin with Sunlight

Oil and coal are made from wood and plants that were crushed under the weight of layers of soil. Gasoline and kerosene come from oil and coal, and alcohol comes from grain and fruit. These can all be used as fuel, fuel that began as sunlight.

Question: What are the main chemicals that you get from burning wood, gas, oil, or coal? (Hint: the same ones you get from the release of energy in respiration. Use the respiration equation to answer this.)

Words to Know

Cellulose is plant fiber, made of a type of glucose.

Chitin is a tough "skin" found on many insects. It is made of a type of glucose.

Words to Know

In the **carbon cycle**, carbon moves from the air, to plants, to animals, and eventually back to the air to be used again and again.

The answer is water and carbon dioxide.

Now, you remember the food cycle. There is also a **carbon cycle**. The carbon cycle begins with photosynthesis, in which plants take carbon dioxide from the air to make food, then animals eat the plants. When the animals die, they decompose and release carbon into the ground. Also, when animals and people breathe, they release carbon dioxide gas into the air as the result of respiration. And now you know that when wood and plants are burned as fuel, they release carbon dioxide back into the air. Plants then use the carbon dioxide released into the ground and air to begin the process of photosynthesis all over again.

You can see some of the carbon cycle by making your own carbon. If not enough oxygen gets to a fire, it produces lots of black smoke, made of carbon. Grey smoke is part carbon, and mostly steam. And, although you cannot see it, there is also carbon dioxide.

To make some pure carbon, have an adult light a candle, and then put a steel pan over the top of the flame. The black soot that collects on the bottom of the pan is pure carbon. It used to be called lampblack from the days when lamps used coal oil or kerosene. This was used to make ink in the pioneer days.

Conclusion

All living things need "fuel" for energy, and to grow. The simple elements—oxygen, carbon, and hydrogen—become larger building blocks when they are joined by bonds. Different things like water and sugars are formed from these elements, depending on the kind and number of elements that are bonded together. But these three simple elements—oxygen, carbon, and hydrogen—all provide fuel for living things.

Roll up Your Sleeves!

Please remember to read all of the activities, right to the end of the chapter, even if you are assigned only one or two activities. Information contained in the activities is also instructional, and part of your lessons!

Activity #1

Draw a picture of a campfire, surrounded by a tree and animals, a boy, grass, and light from the sun, in your notebook (or copy the drawing above). Show the boy cooking something over the campfire. Use arrows to show how everything in creation works together in God's plan.

Label each thing, indicating what it consumes and what it produces from what you have learned in this chapter. Remember that plants use the process of photosynthesis to produce glucose from sunlight, oxygen, and water. Then the plants are eaten by animals and people.

1. What two things in particular do plants produce that is taken in by people?

2. What do people produce that is taken in by the plants?

3. What does the campfire produce?

4. What product in creation was used to make the campfire?

5. Where did the energy that heats the food over the campfire come from *originally*?

> Behold and See 5 Student Workbook:
>
> Worksheets for Chapter 4 begin on page 27.

81

Let's Explore!: Expanding the Lesson

Pick a topic below to research online or in an encyclopedia. Draw a picture, write a paragraph, or tell someone what you learned.

- photosynthesis
- why leaves change color in the fall
- carbon cycle
- cellulose

Activity #2

You know that glucose is a sugar, but there are lots of kinds of sugars. The smaller the sugar is, the sweeter it tastes. It tastes sweet because your tongue is telling you that this fuel is good for energy! (Sucrose is white table sugar; fructose is sugar from fruit. Both taste sweet.)

Sucrose Molecule

Here are two glucose molecules joined together into sucrose. Can you see where they join? How many bonds are on the oxygen? What was removed to make the link between the molecules?

Correct, two hydrogen atoms, one from each glucose molecule, and an oxygen atom.

Super Challenge!

When sugar is joined together into longer chains (called molecules) it loses its sweet taste and becomes starch. Starch is made of sugar, but does not taste sweet; it is the "fuel" that is in bread, rice, corn, and other foods that come from plants.

Here is what starch looks like:

Starch Molecule

[Diagram of a starch molecule showing seven connected glucose ring units arranged with five in a horizontal chain and two branching above and below, each ring containing C, H, O atoms with hydroxyl groups.]

Questions:

How many oxygen atoms would be needed to make energy from this starch molecule by breaking it down into carbon dioxide and water? How many water and carbon dioxide molecules would be made from this?

For your answer, draw this molecule. Draw circles around the CO_2 and H_2O molecules, then add extra oxygen atoms to use up all the C's and H's.

Do you think this could be used as a fuel for cars? The answer is yes. Starch can be made into fuel alcohol, which can be put in the gas tank of cars that have a special carburetor to make sure enough oxygen gets to the fuel. And when our bodies want to store this carbohydrate, or starch, the glucose is converted to longer chains and becomes fat! Fat is just a kind of glucose put together in long chains which our bodies store to use later as fuel.

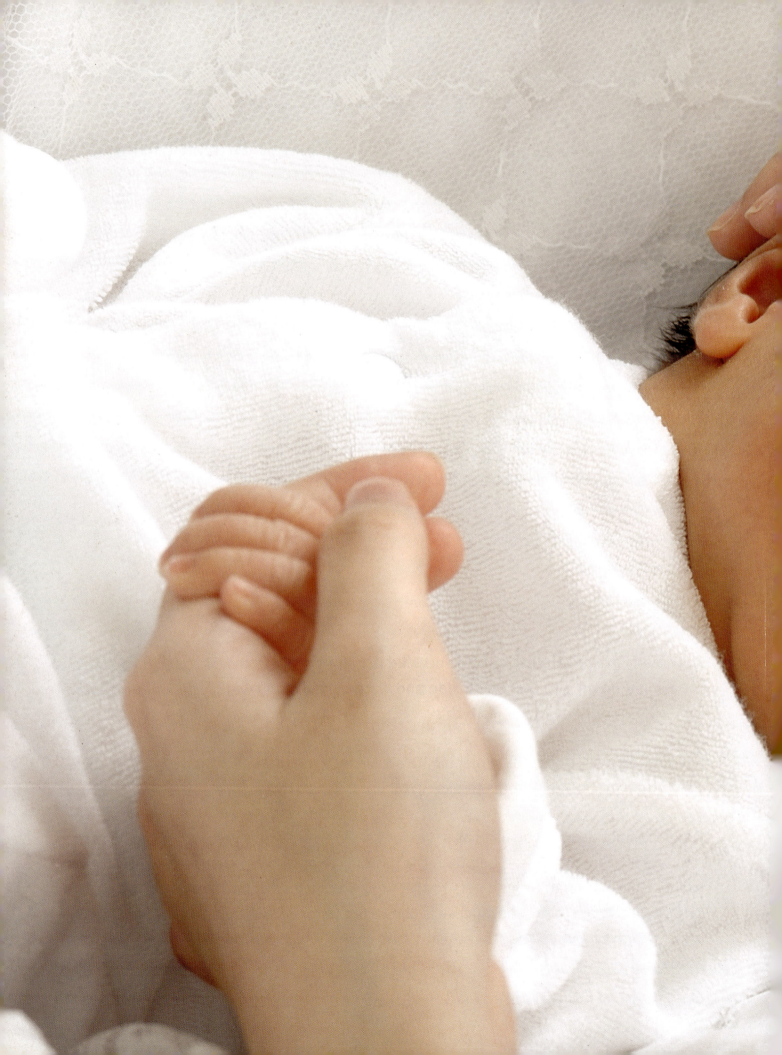

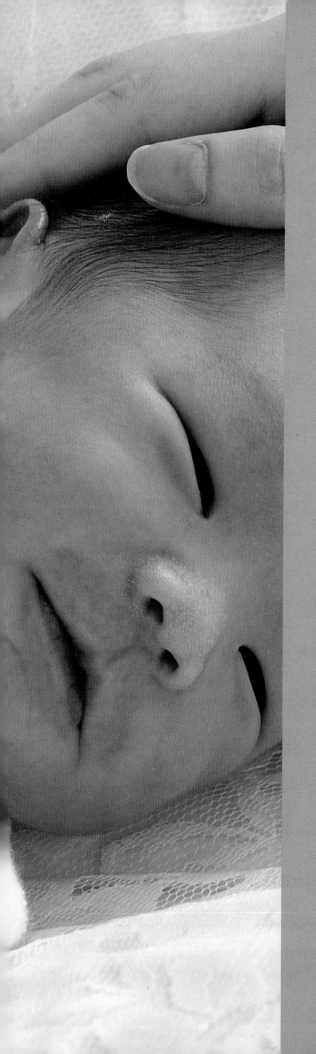

Chapter 5

The Circulatory System and Human Physiology

What is man, that thou art mindful of him,
and the son of man that thou dost care for him?
— Psalm 8:4

The Circulatory System and Human Physiology

Have you ever competed in track or in a foot race? If so, when the race was over, did your heart continue to race while you caught your breath? You gave your **circulatory system** a real workout!

The circulatory system is the body system that circulates oxygen and nutrients from the heart through the body, then back to the heart and **lungs** to remove waste products and begin the process all over again.

This process begins when you breathe air into your lungs, or **inhale**. The oxygen then travels from your lungs to your heart, which pumps it to every part of your body. When the oxygen has been used up, you breathe out the waste products through your lungs, or **exhale**. (Do you remember that respiration is the process of using oxygen to release energy? Respiration is also another name for breathing in and out.)

Can you check your pulse? It is not an easy thing to do. It takes practice to do this well. The pulse is caused by the heart pumping blood, which carries oxygen and fuel to our muscles and bones and anywhere else that energy is needed.

The pulse you feel is the blood being pumped through your **arteries**. Arteries are the hollow tubes called blood vessels that send blood from the heart to the rest of the body. Some people mistakenly call these veins. Arteries are not veins!

Veins are blood vessels that carry "used" blood back to the heart. If you cut your vein accidentally, the blood is dark red and it dribbles out slowly, because it isn't being pushed forcefully

Words to Know

Human physiology is the study of the cells, organs, and functions of the human body.

The **circulatory system** is a system of the body that circulates oxygen and nutrients through the body.

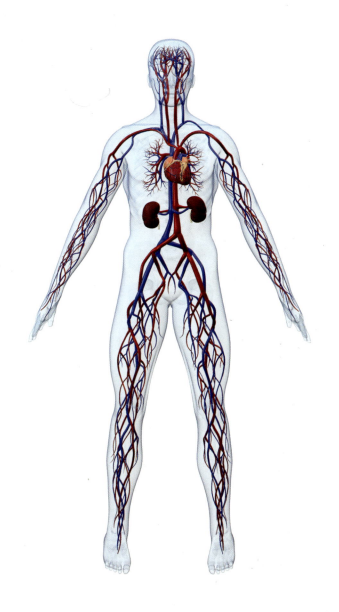

from the heart like the blood in the arteries is. If you cut an artery, the blood is bright red and it squirts out with each pump of the heart.

Turn your hand palm up and look at your wrist. There are blue lines crisscrossing your wrist; these are the veins. If you look closely, you can see them branching out toward your hand.

The blood in our veins is carrying old "used" blood back to the heart and lungs to be cleaned and "restocked" with nutrients and sent out again throughout the body. The blood vessels that pulse

Words to Know

Lungs are organs that fill with air when we breathe in, and send away carbon dioxide when we breathe out.

To **inhale** means to breathe in.

To **exhale** means to breathe out.

Arteries are hollow tubes called blood vessels that send blood from the heart to the rest of the body.

Veins are blood vessels that carry "used" blood back to the heart.

are the arteries, because they are carrying blood full of oxygen directly from the heart.

Now, for the pulse. To take your pulse, you need to hold out your palm, and then place your index finger across your wrist, right below your hand. Do not use your thumb, because there is a pulse in your thumb also. You need to use your finger to feel the wrist pulse.

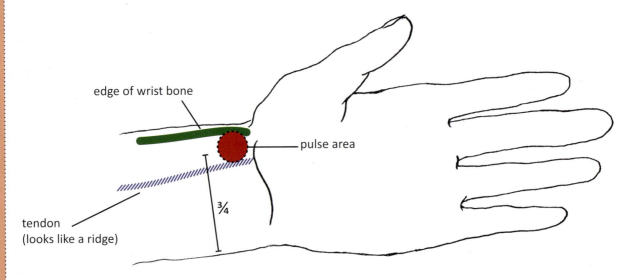

Find the gap between the tendon and the wrist bone on the inside, or outer edge, of your wrist. The pulse is in this gap. If you put a penny on your wrist so the edge of the penny is touching the edge of your wrist, the pulse will be under the middle of the penny.

Count how many beats there are for 1 minute. Now, do it again, but only count how many beats there are for 15 seconds and multiply this by 4 to get the number of beats per minute. Are your answers the same? They should be close. We will discuss your pulse more later on, so it is a good idea to practice finding it now.

The blood in the arteries carries oxygen and nutrients from the heart to where it is needed. The arteries are just like the gas lines in a car which is pushed by a fuel pump to the engine. The fuel pump in our body is the heart, but what is the fuel that our hearts pump?

It is glucose, the sugar that you have learned about in previous chapters.

The red color of our blood is caused by an iron pigment (it is red, just like rust is red) called **hemoglobin**. This pigment carries the oxygen we breathe with our lungs throughout our bodies. The hemoglobin attaches itself to the oxygen in the lungs and carries the oxygen through the arteries to wherever our body needs it. The oxygen is used for combustion, or "burning," of the fuel glucose that gives our muscles energy.

Once the oxygen has been used to release energy in our muscles, the "waste products," carbon dioxide and water are left. (Remember the chemical equation for respiration in the last chapter? H_2O (water), CO_2 (carbon dioxide), and energy were the results of the process, on the right side of the equation.) The blood carries these "waste products" away from our muscles. The hemoglobin attaches itself to the carbon dioxide (CO_2) and carries it through our veins to our heart and lungs.

In the lungs, the CO_2 and extra water leave the blood cells as we breathe out. When we breathe in, a fresh supply of oxygen is attached to the hemoglobin in the red blood cells and the process starts all over again. This is God's design for feeding, or fueling, our bodies.

In summary, blood carries energy and oxygen to your muscles, and then brings back the carbon dioxide and waste products to

Words to Know

Hemoglobin is the pigment in blood that carries oxygen through our bodies.

be cleaned. The carbon dioxide leaves your body from your lungs when you breathe out. Then oxygen goes into your blood when you breathe in.

You have learned that the blood vessels that go out from your heart are called arteries, and the blood vessels that go into your heart are called veins.

Now we will learn a little more about that very special muscle and pump, the heart. The heart has four parts or chambers. The upper chambers are the right **atrium** and left **atrium**. (When we mention more than one atrium, we say the plural **atria**.) The lower chambers of the heart are the right and left **ventricles**.

Have a look at the diagram on the opposite page to see how the heart works. Notice the way the used blood flows from the veins into the right atrium of the heart, and then is pumped through the right ventricle up toward the lungs. In the lungs, the CO_2 is removed, and oxygen is picked up.

Then fresh blood from the lungs, full of oxygen, flows into the left atrium. From there it is pumped to the left ventricle, then out to the body. (Notice that, since we are looking at the heart from the front, left and right are reversed.)

Words to Know

An **atrium** is an upper chamber of the heart.

A **ventricle** is a lower chamber of the heart.

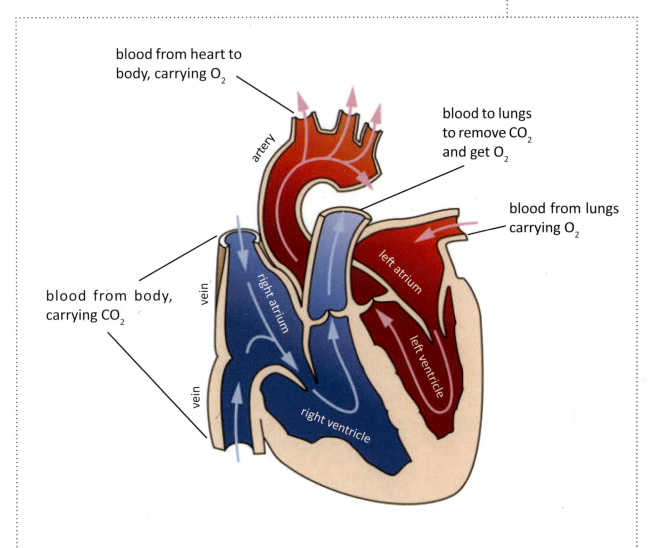

Cross-section of heart, showing the valves in each chamber of the heart.

Words to Know

Emphysema is a disease that destroys the lungs.

Now look at the heart in the photograph below. This is from a lamb our neighbor raised, and we used it to have a lamb roast. (I kept some of the organs to see if there were any parasites in them, but they were not infected.)

The white material on top of the heart is fat. (Did you remember that fat is a kind of glucose put together in long chains, and used for fuel storage?) This heavy kind of fat on the heart is called tallow. (Tallow is used in some recipes for mincemeat, and also was used not so long ago to make candles. Tallow can also be used to make soap by mixing it with wood ashes soaked in water.)

After looking at the photograph and studying the drawing of the heart, do you see how marvelously God designed the circulatory system? It is this circulatory system that makes our hearts and lungs and blood move oxygen and nutrients throughout our bodies all day, every day, even when we sleep. "In Him we live and move and have our being." (Acts 17:28)

Heart and Lung Health

If we fall and scrape a knee, it is easy to see that there has been damage done to our skin! But because our circulatory system—heart, lungs, and blood vessels—is inside us, it isn't so easy to know when it has been damaged.

So, what kind of damage could happen to the circulatory system without our being able to see it, and how can damage be prevented?

The heart is a muscle, and muscles are made strong by exercise. So one of the first ways to keep the heart strong is to exercise regularly. If possible, every day should include exercise like hiking, biking, dancing, swimming, roller skating, playing tag, walking a mile to Holy Mass every morning, jumping rope, or some other kind of active sport.

A second way to keep the heart healthy is to eat a nutritious diet, rich in fruits, vegetables, whole grains, and good quality protein, but reasonably low in fatty junk foods and animal fat. Too much animal fat can cause arteries to become narrow, so blood has a harder time flowing to get to your muscles. Then the heart has to work harder to push the blood through those narrow arteries to your muscles. And narrow arteries can sometimes get plugged so the blood can't flow at all in the area of the "plug," which can cause a heart attack.

A third way to keep both heart and lungs healthy is never to smoke. Smoking can damage both the heart and lungs. One type of lung damage is called **emphysema**, a horrible disease that cannot be cured. Emphysema gradually destroys the lungs, bit by bit, so that the person eventually cannot breathe!

The way God made our circulatory systems work so well is simply astounding. It is up to us to *keep* this astounding system working well!

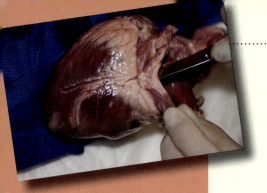

Let's put what we've learned to work.

Roll up Your Sleeves!

Behold and See 5 Student Workbook:

Worksheets for Chapter 5 begin on page 34.

Please remember to read all of the activities, right to the end of the chapter, even if you are assigned only one or two activities. Information contained in the activities is also instructional, and part of your lessons!

Lab Activity #1

THIS ACTIVITY MUST BE DONE WITH AN ADULT BECAUSE IT INVOLVES THE USE OF A KNIFE. DO NOT DO THIS ACTIVITY WITHOUT ADULT ASSISTANCE AND SUPERVISION.

From the meat department in a grocery store, or from a butcher shop, purchase two beef or lamb hearts. It is best to purchase two, as you will have an opportunity to examine the heart from a second angle.

You will need a knife for cutting and a cutting board set on a stable surface. You may also wish to use a camera to take pictures at each step of your lab activity.

First, put one heart in the kitchen sink. Examine the heart. What do you see? Is there tallow, or fat, on the heart? What does the tallow's presence or absence tell you about the diet of the animal from which the heart came?

Now look for blood vessels, or openings where main vessels enter and exit the heart. Do you see any other, smaller blood vessels?

Turn on the tap and run a little water into the heart. Gently squeeze the heart. Can you see how the blood might move through the heart?

94

Now set the heart on the cutting board. DO NOT CUT THE HEART WITHOUT AN ADULT'S SUPERVISION. Dissect, or cut apart to study, the first heart. Can you locate the ventricles at the bottom of the heart, and the atria at the top of the heart? Now dissect and examine the second heart from a different angle and see if you can identify the ventricles and atria again.

Now that you have performed your first dissection, you may wish to write about your experience. If you have taken photos, you may include them in your report.

Activity #2

Using the diagram of the heart from this chapter as a model, draw the heart and its atria and ventricles. Label the atria and ventricles. Add arrows that show the direction of blood flow. Be sure to write in where the blood is headed, and what chemicals or nutrients it is carrying.

Activity #3

Draw the vein patterns of your left and right wrists in your notebook. You can begin by tracing the outline of your hand, and then using a blue colored pencil for the veins. The patterns in each wrist should be similar, but need not be identical (mine are a lot different from each other).

Chapter 6

Logic: Deduction, Induction, and Scientific Reasoning

Man is a noble being created to the image and likeness of God, not insofar as he is housed in a mortal body, but in that he is superior to brute beast because of the gift of a rational soul.

— St. Augustine

Logic: Deduction and Scientific Reasoning

Logic and Reasoning

We have talked a little about the gift that God gave to man alone, the gift of reasoning, or being able to think at a level that animals cannot. Reason is one way in which we are made in God's image. In this chapter, we will learn a little about **logic**, a special type of reasoning. Then we'll use logic to see how scientific questions are based on an **hypothesis** (which is an educated guess about why something happens) and its **predictions** (which is what we think might happen in the future if our hypothesis is correct). We will use the relationship between heart rate, breathing rate, and exercise to see how logic, hypotheses, and predictions are related to scientific research.

First, we will look at the two kinds of logic used in science, deduction and induction.

Deduction

Here is a question:

1. As you know, all triangles have three sides, and no other shapes have exactly three. If I cut some shapes out of wood at my house, which of my pieces of wood has three sides?

a) the wooden square

b) the wooden rectangle

c) the wooden triangle

d) the wooden circle

Words to Know

Logic is a special way of thinking or reasoning.

An **hypothesis** is an educated guess about why something happens based on what we already know and can observe.

A **prediction** is what we think might happen in the future if our hypothesis is correct.

This seems easy, and I am sure you got the right answer *even though you have never seen the pieces of wood at my house!*

Here are some more questions. There is only one correct answer to each question.

2. All goats have no teeth in their bottom jaw. At my farm I have lots of animals, but only one kind has no teeth in its bottom jaw. Which of my animals has no teeth in its bottom jaw?

a) my dog
b) my goats
c) my donkeys
d) my pigs

3. All female honey bees have stingers. I only have one kind of insect with stingers in my collection. In my insect collection, which of the insects have stingers?

a) the ground beetles
b) the dragonflies
c) the female honey bees
d) the butterflies

Words to Know

Deduction, or **deductive logic,** uses general ideas or facts to explain a specific case.

Induction, or **inductive logic,** examines many experiences to figure out what might happen most of the time.

4. When the cooling fluid in a car leaks out, it causes the engine to overheat and crack the engine block. Last year my car overheated, and the engine block cracked. Which problem caused this to happen in my car?

a) The cooling fluid leaked out.

b) The brake fluid ran out.

c) I ran out of gas.

d) The clutch wore down.

If you got these right, then you are already an expert at reasoning using logic. The kind of logic you used is called **deductive logic**, and you *deduced* the right answer. **Deduction** uses general ideas or facts to explain each individual case. You have never seen my insect collection, my pets, or the wooden blocks at my house, yet you know lots about them because you are an expert at deductive logic and can reason.

This ability to reason is one of the things that makes human beings so unique in the world. Animals might be able to think, and they are sometimes very clever, but they are not capable of reasoning on the same level as humans. My dog knows when I am not home, and that is the time he uses to get food from the counter, and I can never catch him at it. A friend of mine has a donkey that can open a gate by removing the chain and using the latch. If you have ever tried to catch a dragonfly in an insect net, you know that you get only one chance. If you miss, the dragonfly stays far away; it has learned to avoid the net.

In Europe in the 1950's, birds learned how to open milk bottles that were delivered to people on their porches in the early morning. The homeowners would come out to the porch and find that the bottle was open and the cream was all gone! The birds taught this to other birds, and soon this was happening all over Europe.

However, as far as I know nobody has found a single example of an animal that can reason using logical deduction. Although animals can learn from their experiences, they cannot understand general ideas or facts. For instance, my dog knows how to steal food from the counter but he does not know that food is good for him because he does not understand how his body converts food into energy. I do not think we ever will find an animal that can reason using logical deduction, because as clever as animals are, they are not human beings made in the image of God and given the gift of reasoning.

Induction

There is another kind of logic called **induction**, or **inductive logic**. Here is how it works.

For years and years I hunted deer, and I found that none of the deer I caught had canine teeth in their upper jaws. (The canines are the sharp teeth on the sides at the front that look like fangs. Most meat-eaters, like people and dogs, have canine teeth.)

Can you make any conclusions about deer from this? Do you think that this might be true of all deer? Any conclusion about all deer having no canine teeth would be based on the seven deer I have caught.

Surprisingly, two years ago I got a deer that did have canine teeth. What can we conclude from this? Right! Most deer have no canines, but there are some that do.

I researched this question and found that less than 1% of the deer in the northeast have canine teeth. So we can conclude that most deer do not have canines, but there are exceptions. However, we cannot be sure how many exceptions there are. We can also conclude that deer, who don't have the ability to reason by induction or deduction, don't care whether they or other deer have canine teeth or not! Man alone would be interested in the science behind this discovery.

So this is how induction works. We look at many individual events or things to see if we can find a general pattern that usually happens. However, we must always remember that we might find exceptions to the rule; we can never be certain of our conclusions the way we can with deduction.

Another example of induction: My chain saw broke, and I phoned the mechanic. He asked me if the firewood I was cutting was wet. I told him yes, it was wet; it had been raining the day before I cut the wood. He said that from his experience, the chain saw was probably just clogged up with wet sawdust. I checked, and he was right. I was able to fix my chain saw by cleaning out the wet sawdust.

Which kind of logic was the chain saw mechanic using?

Correct. He was using induction. Even though he couldn't be absolutely certain, he was telling me the most likely answer from his general experience of fixing chain saws.

Summary of Logic, Deduction, and Induction

Deduction uses a general principle or rule, such as the fact that all triangles have three sides, to understand each individual case. If an object has three sides, we can deduce with certainty that it must be a triangle.

Induction goes from the experience of individual cases to generalization about most cases. Of the several deer I have caught, only one had canine teeth. Therefore, we can induce that most deer don't have canine teeth, although there might be exceptions to the rule.

Notice something about the difference between induction and deduction? Which has exceptions? Can you have an exception using deduction? No, you cannot. Do you see why? If all triangles have three sides, how can there be an exception? There cannot be a four-sided triangle.

With induction, which is just learning from what usually happens, there can be exceptions.

The ability to think logically and to reason are gifts that help us advance in science and other fields by learning more about the world around us. They are also gifts that separate us from all other animals. Let's practice using our logic and reasoning skills!

Please remember to read all of the activities, right to the end of the chapter, even if you are assigned only one or two activities. Information contained in the activities is also instructional, and part of your lessons!

Behold and See 5 Student Workbook:

Worksheets for Chapter 6 begin on page 43.

Let's put what we've learned to work.

Roll up Your Sleeves!

Activity #1

Experiment in Physiology:

You remember from the last chapter that the heart pumps oxygen and glucose, or fuel, to our bodies. When we exercise, we use up fuel. Do you remember what our bodies need to process the glucose-fuel? It's oxygen.

So if we exercise, do you think it will change how fast our hearts pump? How about how fast we breathe? Faster or slower?

When we do experiments, we need to present our findings to others. We need to let people know the purpose of the experiment, and what our findings were. So now we will learn how to present the scientific data, or information, about our experiment. We will begin with an educated guess that explains what we think happens or might happen, which is also called an hypothesis.

Hypothesis (Explaining Statement):

When someone exercises, the muscles use glucose and oxygen that are in the blood to make energy, and then release carbon dioxide to the blood to be removed.

Of course, you learned about the circulatory system in the previous chapters, so you know our hypothesis is true. But there was a time when people did not know how the circulatory system worked. They had to discover the truth by making educated guesses, or hypotheses, and then testing them with experiments. It is thanks to these experiments that we know about the circulatory system today.

Now we are going to test our hypothesis with an experiment. Before we test our hypothesis, we need to write out our expected outcomes, called the predictions. Here are my predictions, based on the hypothesis above.

Predictions: *If our explanation is right, then after someone exercises, the pulse will quicken to send more glucose and oxygen to muscles, and the person exercising will take more breaths per minute. If our explanation is wrong, then after someone exercises, the pulse will stay the same or drop, and his breathing rate will stay the same or drop.*

Question: What kind of logic is being used here to get the predictions from the hypothesis?

Remember that there are two types of logic: induction goes from observation of what is going on to a general idea, and deduction uses a general statement or law to look at particular examples.

So what kind of logic is used in our hypothesis? That's right, deduction!

Here is a chart of what we are going to do. (Please write this out in your notebook!)

> **Hypothesis:** Physical exercise uses up glucose and oxygen in the blood and adds carbon dioxide to the blood.
>
> **Predictions:** If true, after exercise the pulse goes up and breathing rate goes up.
> If not true, after exercise pulse and breathing stay the same or go down.
>
> **Experiment:** Have someone exercise, checking pulse and breathing before and afterward.
>
> **Conclusion:** Did the evidence support the hypothesis or not?

For this experiment, you will need to record your pulse rate and breathing rate while you are resting. Take your pulse for 15 seconds and write this number in your notebook. Do this once every minute for five minutes. Calculate the average of your results by adding them together and dividing by five. This is your normal, or base, pulse.

Now count your breaths for 15 seconds and record this number in your notebook. Do this once every minute for five minutes. Calculate the average of your results to get your normal breathing rate.

Now decide on an exercise. You can choose push-ups, jumping jacks, sit-ups, running in place, or anything that is physical exercise. (I usually do an odd exercise called burpees, which is a push-up followed by a jumping-jack, then another push-up, etc. Ten burpees are a lot for me.) Then take your pulse and breathing rates again. This time you should do them at the same time, by taking your pulse for 15 seconds, using another 15 seconds to write down the result, and then counting your breaths for 15 seconds. Then use the last 15 seconds to write down your breathing rate, so you can be ready to take your pulse again when the next minute begins. Do this for five minutes, or until your pulse and breathing rates go back down to their normal levels.

After your experiment is finished, **multiply each of the numbers by 4** to get your heart rate and breathing rate per minute. Set up your data table in your notebook to look like the one below.

You can make this experiment better by including more people, so try to get your family or friends to do it. You can try this during the morning, and again after eating supper. How do the results change?

Data Table

Name of researcher and test subject _____
Date and time _____
Place _____

Trial	Time	Pulse Count	Breath Count	Pulse/Minute	Breath/Minute
Before Experiment		15 seconds		Per minute (pulse times 4, breath times 4)	
1		21	6	84	24
2		21	5	84	20
3		23	7	92	28
4		22	6	80	24
5		20	6	80	24
Average		21	6	84	24
Exercise					
6		34	10	132	40
7		27	9	108	36
8		25	7	100	28
9		24	6	96	24
10		23	6	92	24

Activity #2

Graph the results of the experiment in Activity #1. Make one bar graph for the heart rate, and one bar graph for the breath rate. Only use the numbers for pulse and breathing rate *per minute*.

For the heart rate graph, the y-axis (vertical) should be labeled "Beats per minute." For the breath rate graph, it should be labeled "Breaths per minute." The x-axis (horizontal) should number from 1-10 the different pulses and breath rates you got from the experiment.

The completed graphs should look like the ones to the right.

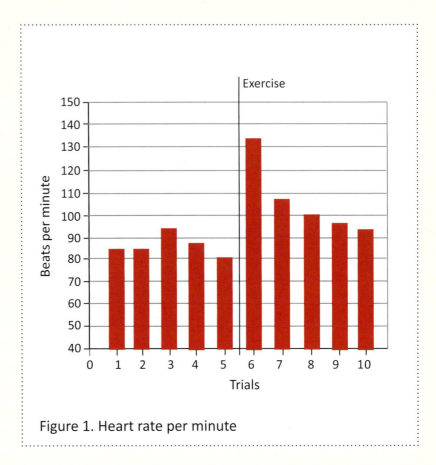

Figure 1. Heart rate per minute

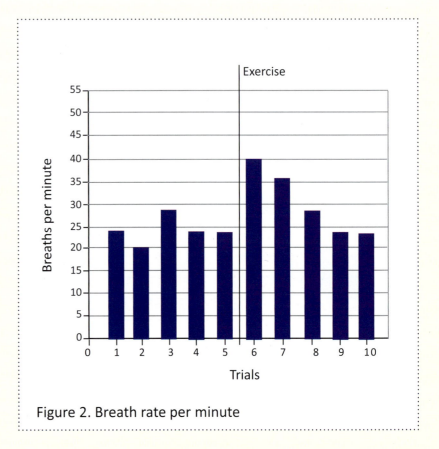

Figure 2. Breath rate per minute

Activity #3: Writing a Scientific Report

You are going to write a scientific report on the experiment in Activity #1. Your report will have sections called *Introduction, Methods and Materials, Figures and Tables, Results, Discussion, References*.

Here is what each section should contain.

Title
What are you studying? (pulse and breathing rate and exercise)

Introduction
- first sentence: what question are you asking?
- second sentence: what is your hypothesis?
- third sentence: what is your prediction?
- fourth sentence: why do you think this?

Methods and Materials
This is just a few sentences telling what you did and how you did it. Write these in the past tense, such as "I took my pulse five times before exercising."

Explain how your experiment was done. Do not use a list.

For this section of your report, draw a picture of where the pulse is. Label this picture Figure 1, and include a sentence under the picture describing what the picture represents.

Figures and Tables
Put your graphs and any tables of numbers in this section of your report. Figures should be in order. Do not tell someone to look at Figure 2 before Figure 1.

Tables work the same way. The first one that you mention is Table 1. Figures have a sentence at the bottom, and tables have a sentence at the top. Here is an example of each, based on an experiment I did with horse flies, showing where to put the sentences.

Figure 3. Trap for horse flies made out of plastic.

Table 1. Number of horse flies and deer flies caught on each day.

Day	Horse flies	Deer flies	Total catch
1	24	18	42
2	26	14	40
3	18	12	30
4	14	7	21
5	27	16	43

Results

In this section, include a sentence that says something like this: "I found that the pulse peaked at 130 beats per minute (Figure 1)."

See how this works? The reader will then look at your graphs. After presenting what you found, tell the reader where to look for this information.

Discussion

Discuss your findings in this section. What do you conclude? Were you right or wrong about your predictions? Was your hypothesis wrong, or was the hypothesis supported by the results of your experiments?

References

On the last page of your report, mention the title and author of this book, and titles and authors from any other books or research that you might have used to write your report.

That is it! Every scientific report has this basic format. Be sure to draw lots of pictures and graphs; that is what makes it interesting!

Make sure you keep your report. That way you can build on your knowledge. And that is what science is all about!

Finally, give your report to someone to read. In science, it is important to share your discoveries with others.

Chapter 7

Competition among Plants and Animals

We do not know God in His essence but by the grandeur of His creation and the action of His providence.

— St. Maximus the Confessor

Competition among Plants and Animals

Words to Know

To **compete** means to do your best to win something that someone else also wants.

Note to parents and students alike: This chapter deals with the observation of plants. If you live in a colder climate where plants are dormant or covered with snow at this time and therefore cannot be observed well, this chapter may be set aside and studied toward the end of spring.

Competition

In any game, the whole idea is to **compete**, or to do your best to win something that someone else also wants.

Of course, for human beings, just playing the game or being in the competition is winning in a way. If winning first place were the only goal of competition, then nobody would play because there can only be one winner. Competing is fun because it makes us stronger and better at what we do, and can even give glory to God as we use and develop the muscles and skills that He gave us. For humans, competing is part of the fun of the game, even if one comes in second, third, or last place.

Animals and plants also compete; this is how they stay quick and healthy. Competition is when two living things compete with each other for some resource. For example, think of two birds competing for a hole in a tree to nest in. Some holes are better for nesting than others; the birds compete over the best ones, with the losers getting the second-rate holes for a nest or maybe no nesting place at all. The competition may include fighting for the nest, or it can be something as simple as finding the best nest first.

This kind of competition is called **contest competition**, in which the winner takes all and the loser gets nothing.

Other examples of contest competition would be two garter snakes competing for a toad which only one can eat, or wolves competing to be the head wolf of the pack, called the alpha wolf. There can only be one of these "bosses" in a wolf pack. The other wolves in the pack must fall into line or be forced out of the pack. It is a contest in which there is only one winner.

The other kind of competition is called **scramble competition**.

The best way to think of scramble competition is to imagine an old game that my grandfather used to play when he was a young man in the 1920's. He would fill his pockets with pennies. Then, when he saw his nephews and nieces he would toss the pennies in the air and call out "Scramble!" and then enjoy watching as the children raced around gathering up the pennies to keep. Remember that pennies were worth a lot back then!

Examples of scramble competition include a herd of white-tailed deer all looking for the best leaves to browse on, mosquitoes competing for the best place to bite someone where they cannot be easily slapped, or plants using their colorful, fragrant flowers to get the most attention from honey bees to pollinate their flowers.

Many kinds of competition fall in between these two extremes, and are part contest and part scramble.

For example, consider how a snapping turtle catches ducklings. The snapping turtles hide in a pond's shallow water, looking up toward the sunlight.

Then, if the turtle sees a small duckling swimming above it, it slowly swims up from below and grabs the duckling by its feet and

Words to Know

A **contest competition** occurs when two creatures compete but only one wins.

A **scramble competition** occurs when two or more creatures compete to get the best or most of something.

pulls it down to eat it. If there are several ducklings and a few snapping turtles, the turtles will be competing in a contest for the first duckling, which will be the easiest and surest catch. After that it is a scramble, because the mother duck will try and get her young out of that pond and the ducklings will be harder to catch. Because the ducks will be more careful now, the second turtle might miss out entirely.

Notice that the mother ducks are also competing with each other for the best places to swim. The good mothers know from experience where the turtles hide; we would expect them to avoid these places, or keep their ducklings close to them.

Facilitation

Biologists used to think that competition explained almost all relationships among plants and animals, but this is not so. Animals and plants are not always competing against each other. They also help each other survive. You remember that this process is called facilitation, a kind of cooperation between different kinds of creatures, plants, animals, and people.

There are lots of examples of facilitation, even though the help isn't always on purpose. You remember that one example of facilitation is bees pollinating flowers so the flowers can make seeds and fruit. The bees facilitate the plants; without bees, the plants could not live. In turn, the plants facilitate the bees by providing them with nectar.

However, each plant also facilitates other plants by allowing the bees to survive. That is, the more bees survive, the more plants they can pollinate, so all plants are helped.

Example 1: cats and starlings

An interesting example of indirect facilitation, or "not on purpose" facilitation, is between house cats, starlings, and sparrows. Starlings and sparrows compete with other birds for the best places to build nests, and they can be quite aggressive in claiming and defending their nesting spots. One might think that cats are harmful for starlings and house sparrows because they catch them for food. However, cats also eat lots of other birds, such as swallows, robins, and bluebirds. When these other birds are eaten by cats, their nests become available for the starlings and sparrows.

How much do cats actually help house sparrows and starlings by eating other kinds of birds? I do not know, but it is a good idea, or hypothesis, to test. In biology, the obvious answer is not always the right one! If cats do provide help in this way, though, then they are facilitating starlings and house sparrows.

Can you think of an experiment to test this idea? First, let's state the hypothesis formally.

Hypothesis: *House cats facilitate house sparrows and starlings by preying on, or catching and eating, their competitors, making new nesting sites available.*

How will we know if this is true? We need to design an experiment that will tell us if the hypothesis is right or wrong.

We could count the number of songbirds nesting in a woodlot, or on a farm, or in a town or city neighborhood. Then, we could let some house cats go in this area for the summer, and count nests the following year to see if there were fewer songbirds nesting, but more sparrows and starlings. If we did this experiment and found *fewer* starlings and sparrows, then we would know that our hypothesis was wrong, at least over a one year experiment.

Let us state this prediction formally.

Prediction: *If the hypothesis is true, then the year after cats are introduced into a new region, there will be more starlings and sparrows. If the hypothesis is not true, then the year after there will be fewer or the same number of starlings and sparrows.*

I really do not know what the right answer is, because I have not done this experiment, nor do I know of anybody who has. We will not know until somebody does this experiment, or another one like it.

Example 2: cats and mice

Here is another way that cats might facilitate other kinds of animals. Everybody knows that cats eat lots of mice and rats. That is why farmers keep cats in their barns, so that the mice will not rob them of all their grain.

But cats also eat snakes, and sometimes they kill weasels. Snakes and weasels can go right down into the holes where the mice and rats live, and catch and eat the mice. So, it is possible that, by killing snakes and weasels, house cats might sometimes actually help mice and rats survive!

Again, I do not know if this is true. Without doing an experiment, anyone who has an opinion is just guessing at the answer, even if that person is a scientist! It might be a good guess, but without an experiment, it is still just a guess.

This is an important point! All good scientists keep reminding each other that they do not know the answers without doing experiments, and *even guesses that seem likely to be true are just guesses.* To know, we need to do experiments.

Example 3: cats, mice, and parasites

We are not finished with cats yet! Here is another example that shows how complicated things can get in nature.

There is a tiny parasite called *Toxoplasma gondii,* or **T. gondii** for short. *T. gondii* infects mice, rats, and cats, and often can infect human beings by accident. About half of all the people in the U.S. have this parasite! (Most people don't even know that they have the parasite, as it seldom harms people.)

The *T. gondii* parasite does not harm cats, either; it just lives in the cat. However, the parasite needs to get to other cats to **reproduce**, or produce offspring. The way it does this is remarkable.

T. gondii passes out of the cats in their dung (cat manure). If a mouse or rat eats some of the dung, the mouse or rat becomes infected by the parasite. Once in the rodent, *T. gondii* goes for the brain and makes the rodent fearless and unthinking. The mice and rats run around more, and go places they would

Words to Know

To **reproduce** means to produce offspring.

T. gondii is a type of parasite.

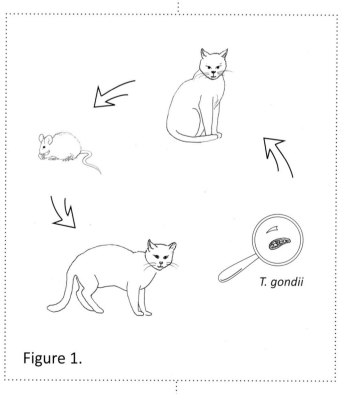

Figure 1.

not normally go during daytime hours. The mice aren't even afraid of cats!

This behavior is bad for the rodent (it gets eaten), but good for the cat (it gets an easy meal) and good for *T. gondii* (it gets into another cat when the cat eats the infected mouse). Cats catch infected rodents easily, and the *T. gondii* parasite infects the new cat to start the process all over again.

Just a moment, let us think about this for a bit. Do the rodents really lose? Can you think of how this might be good for the rodents also? That is, can you come up with an hypothesis for the idea that *T. gondii* parasites benefit rodent populations?

Here is one idea (or hypothesis) that seems to make sense.

Think about how a cat hunts. Like any kind of hunting, it is a learned skill. There are good hunters and bad hunters, and the more practice an animal gets when young, the better hunter it will be.

If there were no infected mice or rats, then cats would have to

develop their skills to be successful, learning where to find healthy mice that are thinking clearly and know how to hide, and when to catch the mice unawares. Catching a healthy mouse that is paying attention is much harder than catching a sick mouse that isn't thinking clearly!

But, if cats keep catching sick mice or rats because they are easier to catch, the cats might not get enough practice to be good at catching healthy ones. Also, they might never learn where and when to hunt for healthy rodents. If a mouse is walking around in plain sight, you don't learn to search all their hiding places, because they aren't hiding.

Now imagine you are a mother mouse who has hidden her pups (yes, baby mice are called pups, too) in a nest under a board or stone beside a barn. If the barn cat keeps hunting and catching sick mice that wander out in the open, far away from your nest, then the parasites would be helping you by keeping the cat away and keeping it out of practice. But if the cat cannot find sick mice and gets hungry, it will begin to look for healthy mice and their nests!

Is this always true? I do not know. (There are countless things that biologists do not know, because there is so much to learn about in God's creation!) It is a good hypothesis that certainly seems reasonable, and is worth testing.

Sowbugs, or pill bugs, can also be infected by parasites. Look at the picture. The sowbug is beside a worm. Sowbugs are actually **crustaceans**, and not bugs at all. This one was under some logs beside my house. Starlings like to eat sowbugs, and when sowbugs are infected, they go out into the open (just like

Words to Know

Crustaceans are animals whose skeletons are made from chitin and are outside their bodies (exoskeleton), similar to a shell.

the infected mice) where starlings can catch them easily. This way the parasite gets passed to another starling. Parasites trick their hosts this way. (Do you remember that a host is a plant or animal that a parasite lives on and feeds on?)

Do these things happen to people? Do parasites cause us to do things that help the parasites? Possibly. Think about how having a cold can make you sneeze on someone. Then that person might catch your cold. Biologists have two hypotheses that might explain this relationship:

1) The cold virus *makes* people sneeze so the virus can be passed on and infect others.

2) A person sneezes to force viruses out of the body so the person can recover faster.

The first hypothesis is that the parasite is **manipulating** us, or in a sneaky way making us do things that we don't plan or want to do.

The second hypothesis is that our body is defending itself against the parasite. Which is the right one? I do not know, and I cannot think of an experiment that would decide which is right and which is wrong.

Most biologists think the right answer is a combination of the two, and that the cold virus is using the body's **defense mechanism**, which is the body's way to protect itself. (For example, you should never look directly at the sun, because this can damage your eyes. But if you look at something in the sky that is close to the sun, you may notice that you suddenly sneeze. This is one of the body's defense mechanisms, because sneezing makes you look away from the sun and protect your eyes.)

Words to Know

Manipulating is a sneaky way of making something do things that it doesn't want or plan to do.

A **defense mechanism** is a way that a body or plant protects itself.

The idea that sneezing when you have a cold could be a combination of different mechanisms seems right to me, but we are not sure. It can be difficult to know if a behavior is a manipulation, or a defense, or a by-product, or some combination of all three.

Competition between Plants

It is easy to find examples of competition with animals, but it is harder to see it with plants unless we know what we are looking for. When we observe plants they seem to just quietly stay where they are and not do much at all.

What we might not realize is that plants are involved in an ongoing competition for everything they use to survive, and that God equipped plants with some remarkable tools and weapons to help them.

Plants compete with each other for lots of things: space, soil, water, sunlight, and nutrients in the soil.

In a woodland or forest, trees grow upward and shade out the plants below them. By spreading out at the canopy area, or the

Words to Know

A **phalanx strategy** occurs when plants grow thickly together to crowd out competitors.

A **guerilla strategy** occurs when plants send out runners that start and connect several plants across an area.

top part of the trees, the trees keep as much sunlight as possible away from the ground so that shrubs and weeds cannot grow very well. This way the trees out-compete the shrubs and weeds for soil nutrients and moisture as well. That is why most older forests have very few weeds or bushes on the forest floor.

Chemical Warfare!

To some plants, our Creator gave characteristics that let them compete by using chemical weapons against their competitors. Black walnut trees send out poison from their roots so that other plants cannot grow nearby. Many different kinds of grass can also do this to stop weeds or broadleaf herbs from growing near them! These poisons don't usually bother humans, but they are effective against other plants.

Phalanx Strategy

Other plants have been given the ability to out-compete their competitors by bunching up in a thick mass, which stops other plants from getting their seeds to the soil or establishing roots. If a seed from another plant does begin to grow in this thick mass, the plants that are bunching together starve the new seedling by taking all the nutrients.

This "bunching together" defense strategy is called a **phalanx strategy**. Now, the word "strategy" suggests that plants are thinking these things out, but of course that is not the case. However, this is the term that is used in science.

Examples of plants who use the phalanx strategy are goldenrod, lilac bushes, cattails, willow shrubs, poplar trees, and most plants used for hedges. The advantage of this bunching, or phalanx strategy, is that it is very hard for any other plants to grow in their midst.

The disadvantage is that plants in a phalanx are so crowded together that they begin competing with each other. Another disadvantage is that if an insect that attacks this kind of plant finds these delicious plants all bunched close together, then all the plants in the phalanx will be attacked by the insect!

For example, in my garden I have red currant bushes which grow in a phalanx. The phalanx is so thick that there are never any weeds growing with the currant bushes.

Three years ago, a wasp called a sawfly found my garden, and in a month all the currant bushes were filled with yellow-spotted, caterpillar-like, immature sawflies that ate all the leaves!

This has happened for three years in a row, because the plants were so thick I was not able to reach through the branches to get the larvae off the leaves. The wasp larvae spread through all the branches, eating as if they were at a free restaurant. I am hoping something that eats sawflies will come along and clear them out for me before the shrubs all die. Plants can't live forever without leaves. Remember how leaves produce food for the plant through photosynthesis? No leaves means no food.

Guerilla Warfare!

A second strategy is called the **guerilla strategy**. Plants using this strategy send out **runners** several inches or even several feet away from the parent plant. Then these runners take root and start new plants away from the parent plant.

This guerilla strategy creates a network of runners that connects several plants across an area. The advantage of this method is that each individual plant is in a different place.

Words to Know

Runners are a kind of sideways stem that grows along the ground away from a parent plant. Then a new baby plant takes root at the end of the runner.

Words to Know

Trichomes are bristly "hairs" that grow on and help defend some plants.

Masting is a cycle by which plants grow fewer seeds for a time and then more seeds in following years.

Strawberries, creeping buttercup, and white clover use this strategy. If some insect or an animal such as a deer or rabbit tries to eat them, not all of them will be eaten because they are so spread out. (If you remember your American history, this was the way the Minutemen fought during the American Revolution against the Red Coats, who used the "bunched-up" phalanx strategy).

Defense Mechanisms Against Animals

Plants also defend themselves from animals using thorns (roses), poison sap (poison ivy), and chemicals. For example, mint leaves are covered in little hairs called **trichomes**, that carry the mint-flavored oil. For us, the mint oil tastes good as a flavoring for food. For insects, the trichomes are miniature spears sticking up from the leaves. The sharp trichomes impale and pierce any insects that try to land on them, injecting the insects with poisonous mint oil. (Poisonous to insects, but not to people.)

Many plants produce poisons to stop animals from eating their leaves. Potatoes and rhubarb use this strategy; their leaves are poisonous, even though the potatoes and rhubarb are not. In some cases, the berries of a plant are poisonous, which keeps animals from eating them. Then the berries can fall on the ground and start new plants without danger of being eaten.

Some plants, especially trees, "attack back" when insects try to feed on the tree's soft inner bark. Examples of this are spruce trees and pine trees. These produce a sticky ooze called sap that flows out of the side of the tree if an insect chews at the bark. The sticky sap from the tree glues the insect by the legs and wings, and then keeps flowing out to cover the entire insect, suffocating it. The ooze then acts as a scab to heal the wound caused by the insect. You can see this on spruce and pine tree trunks. Sap is soft and sticky, and sometimes looks like white chewing gum. (You don't want to get sap on your hands or clothing. It is really tough to get off!)

Masting

Many plants trick their herbivorous (plant-eating) enemies by growing fewer seeds for a time, and then more seeds the following years. This process is called **masting**.

Have you ever noticed that some years there are lots of apples, or fruit, or acorns, and other years there are not very many?

Here is what we think is going on. Consider oak trees. Oaks produce seeds called acorns. Acorns are filled with lots of nutrition, and mice, deer, and squirrels eat acorns. In fact, they eat most of the acorns, as do small beetles and other animals, so there aren't very many acorns left to start new oak trees.

Look at the graph of mice and acorns on the next page (Figure 2). We will use the graph as a model to see how the process works. Let us assume that there is enough food from other seeds to support 500 mice in the woodlot (so the line representing the mice never goes below 500), and that a new oak tree begins to produce acorns on year 1.

Assume that the oak increases acorn production by 5,000 a year until it peaks at top production of 25,000 a year.

Second, assume that 25 acorns will allow a single mouse to survive for a year, and that the number of mice will increase until there is not enough food.

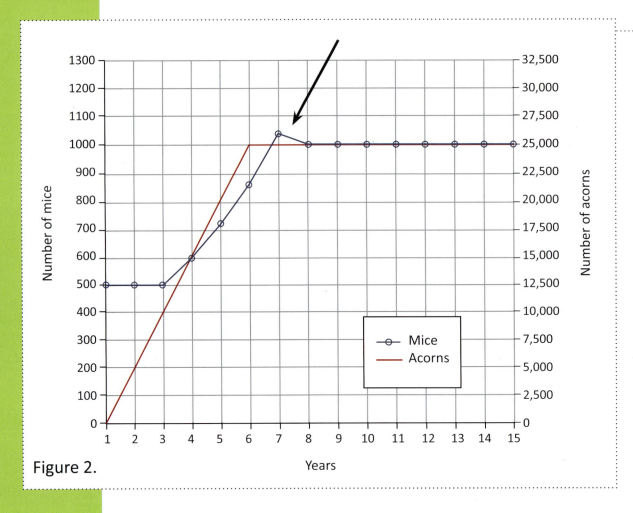

Figure 2.

Eventually the mice will grow so numerous that they will eat all the acorns that the tree can produce and there will be no acorns left to grow into young oak saplings.

Now look at Figure 2 again.

When do the mice numbers start to grow?

When do they stop growing?

Do you see the little blip between year 7 and 8 on the line representing the mice? (The arrow is pointing to this part.) These are the mice that did not survive because there were not enough acorns.

Now, oak trees are not in the business of making acorns just to be kind to mice!

What can the oak do to change things and allow some of its acorns to survive and not be eaten? If the oak stops producing so many acorns, what will happen to the mice?

That's right, some will die. Then the oak can grow more acorns, until the mice catch up. Then the oak will cut back again. The extra acorns will escape being eaten by mice, and these are the ones that might grow into new oak trees.

This is an example of masting, or growing fewer seeds for a time, and then more seeds the following years. Masting is how plants can get ahead of the animals that eat their seeds.

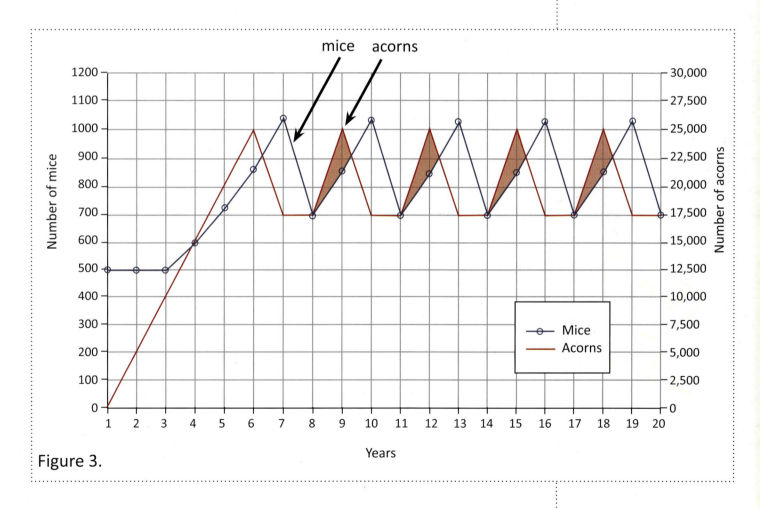

Figure 3.

Look at Figure 3. The shaded part is the extra acorns that can grow before the mice can catch up. This is how trees and plants out-compete the herbivorous animals that attack their seeds.

But it is more complicated than the graphs show, because there are lots of other animals to take into consideration. For example, gypsy moths eat oak leaves, and deer eat acorns, too. It isn't only mice that the oak tree has to defend itself against! But the idea of masting is simple: by using repeating cycles of high numbers of seeds for a time and low numbers for a time, the relationship between two different competitors changes so the plant can survive.

Finally, while some plants compete or defend themselves only with the characteristic of clumping, or sending out runners, or having poison or thorns, other plants possess more than one of these characteristics! For example, some wild roses clump in a phalanx strategy, but are also covered with thorns.

When people compete in games, they do so to have fun and to develop their skills. But competition between plants and animals is a more serious business, because it often determines who flourishes or survives, and who dies. Just like somebody's currant bushes.

Let's put what we've learned to work.

Roll up Your Sleeves!

Please remember to read all of the activities, right to the end of the chapter, even if you are assigned only one or two activities. Information contained in the activities is also instructional, and part of your lessons!

Activity #1

Remember that facilitation happens when two creatures of different species or groups have a working relationship that benefits both groups, like bees pollinating flowers.

Think of two examples that you think might be facilitation, and then think up an experiment that would show if facilitation is at work in those circumstances. You do not need to perform the experiment.

Behold and See 5 Student Workbook: Worksheets for Chapter 7 begin on page 51.

Activity #2

Plant radish seeds in pots according to the numbers listed on each pot in Figure 4.

Figure 4.

Write an hypothesis about whether you think competition for soil nutrients occurs between the radish plants in each pot.

Write a prediction of how you will know if your hypothesis is wrong, or not. Set up the experiment this way:

Hypothesis: Competition _____ occur between radish seeds.   *(does or does not)*

Prediction: If the hypothesis is true, radish plants in pots that are crowded will be _____ than plants in pots that are not crowded.   *(larger or smaller)*

Methods: Plant seeds in pots with potting soil or soil from your yard. Make sure you put the pots in a random order on a tray the way the diagram shows! Also be sure to write the number of seeds planted in each pot as the diagram shows.

Water very lightly every second day, but be careful not to drown the seeds. Just keep the soil damp. When the seeds sprout, then begin to water when the top of the soil feels dry. Do not water so much that the plants are sitting in water; the water should soak in and not be visible on the surface of the soil.

Let the plants grow for four or five weeks. Count the leaves on each plant, and measure how big the leaves are at the end of each week. Also count how many seeds did not grow. (You should have one radish for each seed planted.)

Compare the size of the radishes in the various pots. Note which pots had bigger radishes and which had smaller radishes.

Results: Draw two bar graphs.

In the first one, graph the average number of leaves per plant in each pot. To do this, you should count the number of leaves in each pot, then divide it by the number of plants in the pot. This will give you the average number of leaves per plant in that pot. Graph the different results you get for each pot. Did the plants from the crowded pots grow fewer or more leaves?

In the second graph, compare the average length of the leaves in each pot. To do this, measure the lengths of all the leaves in a single pot, add the numbers together, and then divide by the total number of leaves in the pot. Graph the different results you get from each pot. Were the leaves in crowded pots larger or smaller than leaves in less crowded pots?

Note which pots had the biggest radishes.

Note the number of seeds that did not grow.

Discussion: Did the experiment give evidence that your hypothesis was right or wrong?

Activity #3

You have learned about scramble competition. This experiment involves scramble competition among birds. Really, almost all bird feeders demonstrate scramble competition, but this activity will focus on bluebirds.

Bluebirds are omnivores, but they eat mostly insects. Bluebirds' favorite foods include various insects, larvae, and worms. Since insects tend to lay their eggs in masses, this means that as the eggs hatch lots of larva will be available for a short while before they can spread out to safety under leaves, etc. The birds realize that when larvae are visible, they won't be visible for long. Also, other birds will soon find the larvae, too. So when food is discovered, all the birds scramble!

Bluebirds tend to feed in the early morning, and again in the evening, so that is when you want to put food on your feeder. (And if you don't have bluebirds in your area, you might be surprised at the other varieties of birds that will still show up.)

Dig or purchase about one cup of earthworms or mealworms. (If you don't have earthworms, four hardboiled eggs, chopped, will work.)

Put a piece of plywood or flattened cardboard box at least two feet by two feet on the ground away from the house, where you can still observe from a window. (Worms crawl, so you don't want them to escape into the grass before the birds show up.)

Dump HALF the worms in the center of the board, then go inside and observe.

How many birds came to the feeder?

How many different kinds of birds did you observe?

How long did it take before all the food was gone?

Repeat the experiment again, about half an hour later. Did more birds come this time?

The second part of this activity is building a house for bluebirds, which will keep them coming to your feeder for observation.

After you've built your bluebird house, fasten it to a post or tree about six feet off the ground, and at least 25 feet from a house, facing an open area. (Bluebirds seem to prefer the opening of the birdhouse to face either east or north.)

Any birds will work if there are no bluebirds in the neighborhood.

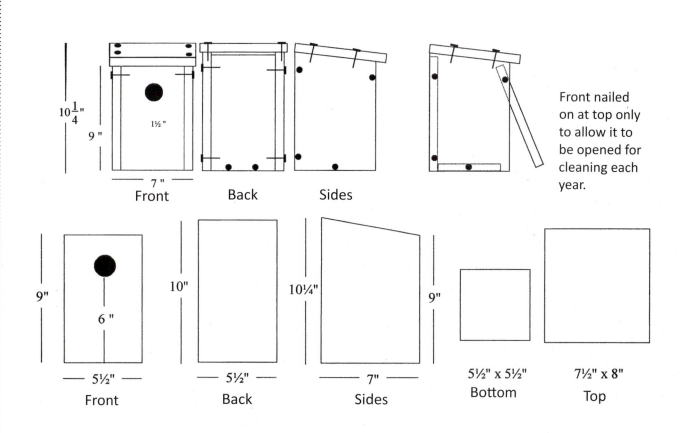

Simple bird house construction suitable for bluebirds or swallows. Requires 56" of 1" (3/4") wood, either pine, cedar, or spruce (NOT pressure treated), and 14 (1/4 lb) 2" ardox or common nails (galvanized).

Chapter 8

Atmosphere and Weather

Not God alone in the still calm we find;
He mounts the storm, and walks upon the wind.
—Alexander Pope

Atmosphere and Weather

In the Psalms, King David marvels over God's majesty witnessed throughout the universe (see Psalm 8). When we study the Earth's atmosphere and weather, we cannot help but be amazed at the complexity of creation and the majesty of its Creator.

Earth's Movements and Seasons

In order to study weather, we have to start with the Earth, because its rotations bring about the seasons. Weather is also determined by where we live on the Earth in relation to the Earth's poles and equator.

The Earth is a large sphere, almost round, but not quite. The Earth is where we spend our entire lives, and it is one of the first solid facts that we experience, and take for granted.

Although it seems stable, in reality, the Earth is a dynamic, changing place that is moving in many directions. It spins like a top; it circles the sun. The Earth and sun together are part of our galaxy, the Milky Way, which in turn rotates like a bicycle wheel.

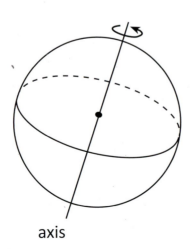

axis

We cannot feel these movements, because we are traveling with the Earth. It is like traveling in a car or train; we do not feel the speed unless we hit a bump, or speed up or slow down. We feel any *change* in speed, but not the traveling speed itself.

Words to Know

The Earth's **axis** is the imaginary center "line" around which the Earth rotates.

The **equator** is the "dividing line" that separates the Earth into two hemispheres, north and south.

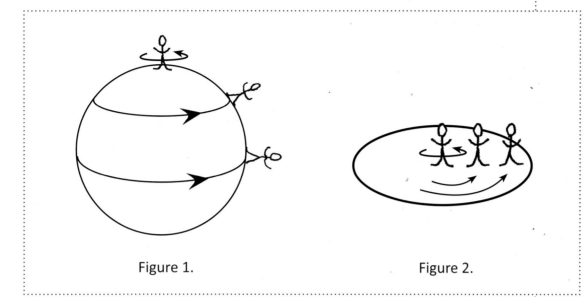

Figure 1.   Figure 2.

How fast does the Earth move? It takes 23 hours, 56 minutes, and 4.09 seconds (or 23.93636 hours) to make a complete rotation or circle around its **axis**, or the imaginary center line around which it spins. That is almost exactly 24 hours, or one day.

This means that someone on the **equator** (which is the dividing line that separates the Earth into two **hemispheres**, north and south) moves around the Earth's axis at a little over 1,040 miles per hour. Because the Earth is a sphere, though, the "spinning" grows slower as you go from the equator to the poles, until at the North Pole and South Pole it approaches 0 miles per hour! (See Figure 1.) Imagine standing on a merry-go-round. Someone at the very edge will move very fast, but someone in the center will only spin in place. (See Figure 2.)

The Earth also travels in an **orbit** around the sun. It takes a little more than a year for the Earth to travel around the sun, or 365 days, 6 hours, 9 minutes, and 9.54 seconds. We add up the extra six hours each year, and correct the calendar with a leap year every 4 years. A leap year has 366 days, with February getting the extra day (February 29). The Earth travels 595 million miles to go around the sun.

Words to Know

A **hemisphere** is half of the Earth's sphere; the Northern Hemisphere is above the equator, and the Southern Hemisphere is below the equator.

The Earth's **orbit** is the "path" it takes, traveling around the sun.

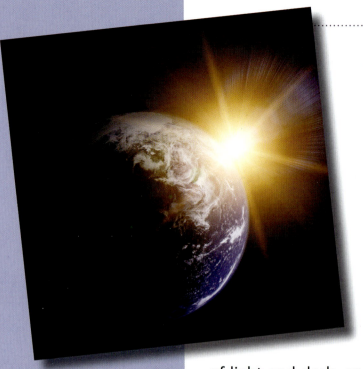

The sun and Earth are part of a huge group of stars, called a galaxy. Our galaxy is the Milky Way, and it also rotates. It takes 200 million years for the Milky Way to make one complete rotation within God's vast universe. So, the Earth does a lot of traveling through space!

Although we cannot feel the Earth move, this movement does affect us. The first thing we notice about the movement of the Earth is that each day has periods of light and dark, or day and night. This is caused by the Earth spinning on its axis so that part of the Earth faces the sun, and part of the Earth faces away from the sun.

Words to Know

The **atmosphere** consists of layers of air and gases that surround and blanket the Earth.

Latitude is a measurement of distance north or south of the equator.

Seasons

The seasons are also caused by the Earth's motion through space. They are caused by the fact that, as the Earth revolves around the sun, its axis is slanted. This slant makes the Earth tilt so that on June 21, the Northern Hemisphere of the Earth tilts toward the sun, and on December 21, the Northern Hemisphere tilts away from the sun.

This tilting of the axis causes the seasons. In the summer, when the Northern Hemisphere is tilted towards the sun, the sunlight goes through the **atmosphere** and shines directly onto that hemisphere. (See Figure 3.)

Figure 3.

In winter, the Northern Hemisphere is tilted away from the sun, so some of the sunlight reflects off the atmosphere. More important, the sunlight is spread out over a larger area because the surface is tilted, so there is not as much heat from the sun on the northern surface. The farther north one goes, the more the sunlight is spread out across the surface, so the temperatures grow increasingly colder. (See Figure 4.)

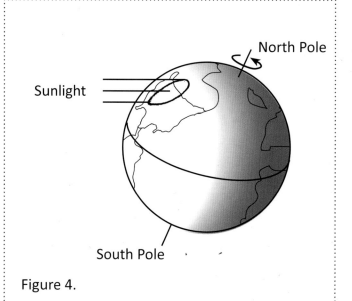

Figure 4.

To understand a little better how distance from the equator influences weather and climate, look at a map or globe and find the lines of **latitude**, which measure the distance north and south of the equator. Lines of latitude north of the equator measure the Northern Hemisphere all the way to the North Pole. Lines of latitude south of the equator measure the Southern Hemisphere all the way to the South Pole. Compare the weather in a country far away from the equator with the weather in a country near the equator.

The tilt of the Earth's axis causes other curious features. Because of the tilt, in the far northern regions of the Northern Hemisphere, children can play baseball in the summer sunshine at 11:45 p.m., then watch the sun set and rise again within a very short period of time. Days are very long and nights are very, very brief. If you look at Figure 5 you can see why. Because the Earth's axis is tilted, a person will spend hardly any time at all in the Earth's shadow as the globe rotates. In winter, though, it is just the opposite, and the days are short while the nights are long.

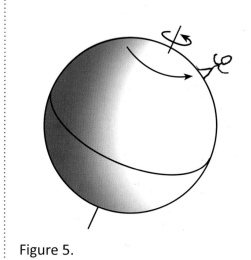

Figure 5.

At the South Pole, the seasons are reversed. There, it is cold and dark during our summer months, and light all day during our winter months. In other words, when regions north of the equator are experiencing a certain season, regions south of the equator are experiencing its opposite. This is because whenever the Northern Hemisphere is tilted towards the sun, the Southern Hemisphere is tilted away from the sun, and vice versa.

On the equator, the length of daylight and darkness stays the same all year round. The sun rises around 7 a.m. and sets around 7 p.m. every day of the year.

Those of us who live about halfway between the equator and the North Pole experience long days and shorter nights in summer. In the winter, we have short days and long nights. Is it the same where you live?

If you would like to watch the cycles of night and day moving across the globe as the Earth rotates on its axis, you might log on to *www.fourmilab.ch/earthview* or do a web search for "satellite image earth at night."

Remember that the Earth is a globe, so it is impossible to see all sides of the Earth at once. The satellite images are put together much like a "flattened" mapped surface, so that viewers can see the entire surface of the Earth at the same time. Because of this "flattening" of the globe for viewing purposes, night's shadow is "flattened" as well.

Solstices

The summer **solstice,** the longest day of the year, is June 21 in the Northern Hemisphere. The winter solstice, the shortest day of the year, is December 21.

At exactly halfway between June 21 (the summer solstice) and December 21 (the winter solstice), the days and nights are equal. Can you figure out why?

That's right, because the Earth's axis is not tilted toward or away from the sun, but to the side so that the day and night are almost the same length of time. Why *almost* the same length of time for both day and night? Because daylight begins just before the sun comes up, and there is still some daylight left even when the sun is just below the horizon. So the daytime is just a bit longer than the night on the two **equinoxes**, or days when night and day are of more or less equal length. ("Equinox" comes from Latin and means "equal night.") These "equal" days occur on the spring equinox (March 20 or 21) and fall equinox (September 21). (See Figure 6.)

Words to Know

The **solstices** are the longest and shortest days of the year.

An **equinox** is a day when daytime and nighttime are equal.

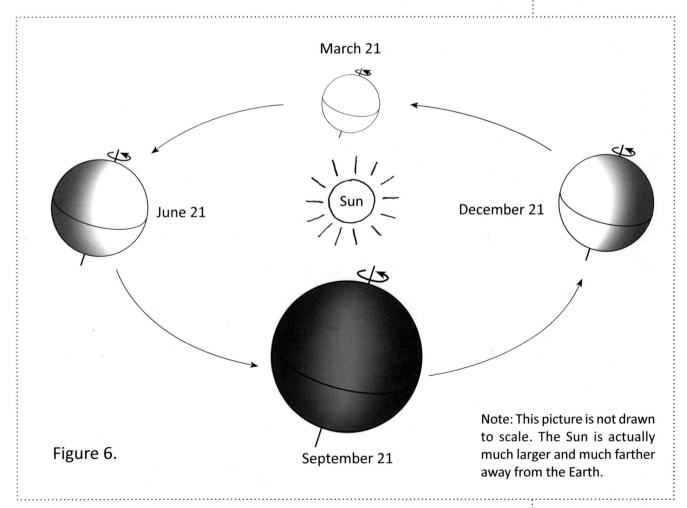

Figure 6.

Note: This picture is not drawn to scale. The Sun is actually much larger and much farther away from the Earth.

Words to Know

The **troposphere** is the atmospheric layer closest to the Earth.

Vapor is another word for gas; water vapor is water in its gaseous form.

The Earth's Atmosphere

The atmosphere is the name for the air and gases above the Earth's surface. This atmosphere is about 300 miles thick, and is divided into layers depending on how high up one goes. These layers don't have exact boundaries, or points of beginning and ending; instead, they gradually blend into one another where the layers meet. We will examine four of these layers.

The air closest to the surface, where everything lives, is called the **troposphere**. The troposphere ranges from about 5 to 12 miles thick, depending on the seasons and also your distance from the poles and the equator. The air in the troposphere is dense; half of all the air molecules in the atmosphere are in the 6 miles closest to the earth. (That is why the air is so thin at the top of Earth's tallest mountain, Mt. Everest, which reaches almost 5 ½ miles into the troposphere, or even the lower boundaries of the stratosphere!)

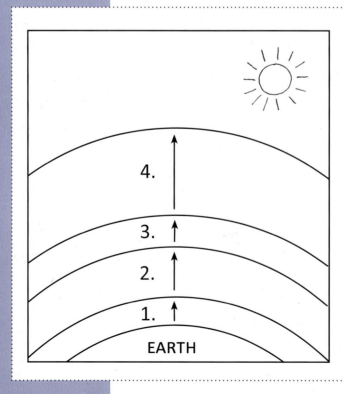

LAYERS OF THE ATMOSPHERE

1. **Troposphere**: begins at Earth's surface, and gradually blends with the stratosphere between 5 to 12 miles above the Earth (10-15 km)

2. **Stratosphere**: extends to about 31 miles (50 km)

3. **Mesosphere**: extends to about 50 miles (80 km)

4. **Thermosphere**: extends to about 300 miles (480 km)

In the lower 5 miles, the air is made up of 78% nitrogen, 21% oxygen, less than 1% argon gas, and 0.037% carbon dioxide, followed by small amounts of neon, helium, methane, krypton, nitrous oxide (laughing gas), hydrogen, and xenon. The troposphere also contains about 90% of all water **vapor**, including clouds. That means a good deal of our weather forms in the troposphere. Most large aircraft try to travel in the upper parts of the troposphere so that they can fly above the clouds and weather systems that make flight more difficult.

The layer above the troposphere is the **stratosphere**. Some aircraft fly in the lower regions of the stratosphere, and the upper parts of some storm clouds can reach this area, too. The stratosphere extends from the top of the troposphere to about 31 miles above the surface of the Earth.

Above the stratosphere is the **mesosphere**, which extends to about 50 miles. The **thermosphere** is the uppermost layer and ends about 300 miles above the Earth's surface. These two layers together form a protective shield against meteorites. Meteorites, also known as shooting stars, are pieces of rock and metal that fall to Earth from outer space. When they pass through the mesosphere and thermosphere, friction makes them burn up until either nothing is left, or the pieces are so tiny that they don't cause harm when they strike the Earth.

The thermosphere is the atmospheric layer into which we launch man-made satellites, and it is also the one in which astronauts in space shuttles orbit Earth. Polar lights are also formed in the thermosphere. These stunningly beautiful displays of "heavenly lights" are called aurora borealis in the Northern Hemisphere and aurora australis in the Southern Hemisphere. Although they form at a tremendous

Words to Know

The **stratosphere** is the second layer of the atmosphere from the Earth's surface.

The **mesosphere** is the third layer of the atmosphere from the Earth's surface.

The fourth layer of the atmosphere from the Earth's surface is the **thermosphere**.

Words to Know

Altitude is the measurement of distance from the surface of the Earth into the atmosphere.

Solar radiation is invisible rays or waves of energy, some harmful, emitted by the sun.

distance from Earth, their brilliant colors can still be seen below! (Unfortunately, with rare exception, these colorful displays are normally visible only in the far north and far south. Can you see the aurora borealis or the aurora australis where you live?)

These upper atmospheric layers, the mesosphere and thermosphere, are very cold because the air is so thin. We cannot breathe the air at these great **altitudes**, or heights, because there is simply not enough oxygen there. In fact, it is even hard to breathe the thin air on mountain tops that jut into the upper half of the troposphere, as all mountain climbers know.

All these layers in the atmosphere protect the Earth from **solar radiation**, which is harmful, invisible rays or waves of energy that come from the sun, far outside our atmosphere. If the Earth were exposed to sunlight directly, without the protective shield formed by the atmosphere, we would quickly die because of the harmful radiation.

Of course, not all solar radiation is dangerous—some of it is the visible light that allows us to see! Long-wave radiation is relatively safe. But short-wave radiation is more intense and powerful than visible light, even though it is invisible. Ultraviolet light, for instance, can give you a sunburn. When sunlight travels through the atmosphere, the atmosphere filters out the more

dangerous short waves, including gamma rays, X-rays, and much of the ultraviolet radiation. The sunlight that then reaches the Earth's surface consists of some ultraviolet radiation, visible light, and long infrared waves, or radio waves.

All these layers of the atmosphere that circle and blanket our planet "belong" to the Earth, for they travel with our planet as it travels around the sun.

Water on the Earth

About 70% of the Earth's surface is covered in water, most of which is in the oceans; the rest of the Earth's surface is made up of the land on which we live. To picture this, think of a giant ball of gum that has been stretched out on a sidewalk in the rain, with an ant crawling its length. For every three inches that the ant crawls on dry gum, it has to cross seven more inches of gum covered with puddles. Puddles cover 70% of the gum's surface, and the ant is very sticky.

The fact that God created water instead of some other liquid to cover so much of our planet's surface is very significant. Water is the only substance that floats when it is a solid, or in its frozen state. This is because water expands, or grows larger, when it becomes a solid, as ice.

All other substances contract, or shrink, when they become solids, and become heavier than they were in their liquid states. Water is actually less heavy as solid ice than it is as liquid water! When water freezes, the ice floats on top of the ocean or lake, instead of sinking. (This is important to fish, because otherwise they would all die in the winter, crushed beneath layers of sinking ice. Isn't God good to take care of the fish by making ice float?)

Words to Know

To **evaporate** means to turn into a vapor or gas.

To **condense** means to turn from a gaseous state to a liquid state.

The **water cycle** is the cycle of evaporation and condensation that provides a never-ending supply of water to the Earth.

Another way that water is different from most other liquids is in the way it heats up. You will find out why this is important as we begin our discussion of weather.

Weather and the Water Cycle

All weather comes from sunlight, and not just on sunny days! What would we do without the sun? We have learned that it provides light, energy for life for plants and animals, and so many other things! (And what would we do without the Son, Who gives Light and Life to all who follow Him?)

Energy from the sun drives the weather. Sunlight heats the water in oceans and lakes, causing it to **evaporate**, or change from its liquid state into a gas, and form clouds. These clouds rise through the atmosphere, because heat rises. To understand this, think about which way steam travels from a boiling pot. Yes, it rises because it is hot. Put a couple of ice cubes in a bowl, and hold your hand about an inch above the bowl. Then put your hand an inch or so below the bowl. Do you feel the cold air flowing more strongly above the bowl, or below the bowl? Warm air rises; cold air sinks. This very simple principle is the basis for most of our weather.

As the warm clouds grow full of water vapor and rise into the cooler air at higher altitudes, they cool. Cold air can't hold as much moisture as warm air, so much of the water vapor in the air **condenses**, or returns to its liquid state, and falls to the ground as rain. The rain soaks into soil, and runs into little creeks and streams, which run into lakes or rivers. When the sun heats the water in oceans, lakes, and puddles, the water evaporates once more and the whole cycle begins again.

This process, in which clouds of warm, evaporated water rise through the atmosphere, and then condense and fall onto land, lakes, and oceans to supply the Earth with a never-ending supply of water, is called the **water cycle**.

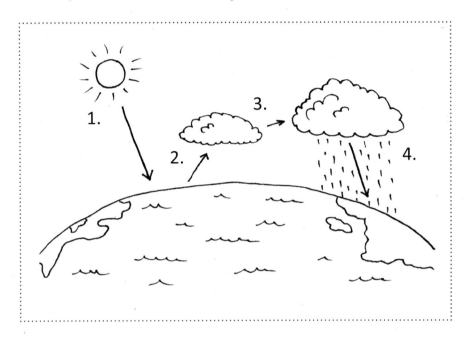

Water Cycle

1. Sunshine warms water over land, oceans, and lakes, causing water to evaporate, or become vapor.

2. Vapor rises with the rising warmer air and forms clouds.

3. Clouds grow larger; cooler air above the Earth cools the vapor in the clouds, causing it to condense, or return to its liquid state.

4. Precipitation falls onto the Earth, watering land, lakes, and oceans, continuing the cycle forever.

This is only half the story, though. In order for hot air to rise, there has to be cooler air to take its place. How does this work?

Once again, it all starts with the sun. When the sun shines on the Earth, some surfaces reflect the light, and some surfaces absorb the light as heat. The darker a surface, the more sunlight it absorbs, and the hotter it gets. (White surfaces do not absorb very much energy; instead, the energy is reflected, or bounces off. This is why black T-shirts are hotter than white ones.)

So dark surfaces, such as asphalt, concrete, and even bare soil, get very hot. How do you think all the asphalt and concrete in the city might influence temperatures in the city? And how might the bare dirt of deserts, with so few growing things, influence temperatures? (Farmland and forests also absorb sunlight and get warm, but not as hot as asphalt or bare soil!)

Bodies of water, on the other hand, usually stay very cool. This is because water allows sunlight to pass through it, which allows energy to be absorbed throughout the entire body of water, not just on the surface as happens on land. Because the sun is heating the entire body of water instead of just the surface, the temperature of the water rises very slowly, and usually stays quite cool. The larger the body of water, the slower the temperature rises. That is why the huge, deep oceans have such a powerful effect on weather.

So the process begins with sunlight warming the ground where there are bare soil, open fields, and cities. The air above this ground heats up and then rises to form clouds. Then the cooler air over lakes and oceans and farmland blows in to fill the space left over by the rising air. As soon as the cool air from the oceans and farms reaches the hot cities, it begins to heat up, and eventually it starts rising, too. More cool air from the oceans and farms blows in to take its place.

Now, the air over water and forests is not only cooler; it is also moister, which means there is more water vapor in it! When the moist, ocean air is heated by the sun and rises, it forms clouds very quickly, and often produces rain. However, the rain does not just fall on the city, because wind often blows the rain clouds right back to the farmlands, forests, and oceans!

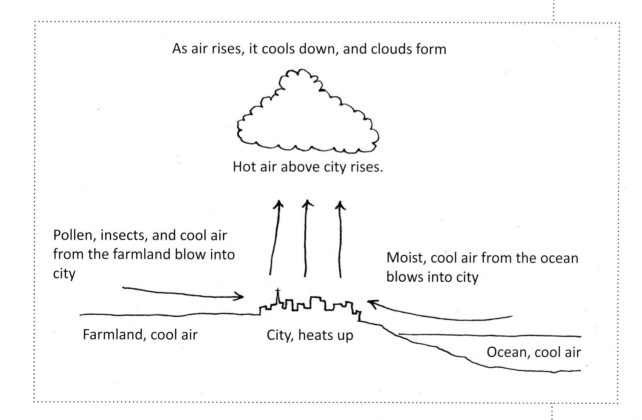

Words to Know

Hurricanes, cyclones, and **typhoons** are all names for severe storms that form over the ocean.

Precipitation is water, in one form or another, falling on the Earth.

Weather from the Oceans

When the water cycle happens over the ocean, it effects not only those who live near the ocean, but also those who live far away. For example, on the continent of North America, heated, rising air in the west often draws cool, moist air from the Pacific Ocean across much of the continent. That is why weather frequently seems to move from west to east across North America.

The water cycle and air temperatures aren't the only causes of weather systems. Water in the tropics (near the equator) is warm, and a warm water current flows from the tropical Gulf of Mexico across the Atlantic Ocean to Europe. At the same time, icy cold water from the Arctic Ocean flows south. The warm, tropical water flows through the cold, arctic water as it crosses the Atlantic. Because the water from these two areas has such different temperatures, air rising over the two currents causes massive clouds and storms to form. Severe storms that begin over the oceans are called **hurricanes**, **cyclones**, or **typhoons**. Since these storms form when there is a great contrast between warm water and warm air versus cold water and cold air, they usually develop between June and November in the northern hemisphere, and between October through March in the southern hemisphere.

152

Types of Clouds

Clouds carry moisture, which falls as **precipitation** in many forms, including rain, hail, sleet, and snow.

Clouds have different shapes and forms, depending on how high they are in the atmosphere, and whether they carry rain or snow. So that we can differentiate between clouds, cloud types have names that tell us more about what they do. For example, *nimbo* or *nimbus* in the cloud name means that it will probably cause precipitation in some form.

You should memorize the types of clouds, and know what kind of weather they are associated with. Now that you know a little about the atmosphere and altitudes, you can see from the cloud chart on the next page where the clouds form in the atmosphere. You can also see why most large aircraft fly high in the troposphere, so they fly in calm sunshine above clouds and storms.

As you learn to recognize the different types of clouds and what kind of weather they may bring, you can begin to forecast the weather!

STRATOCUMULUS

CUMULUS

CUMULONIMBUS

NIMBOSTRATUS

CIRRUS

ALTOSTRATUS

ALTOCUMULUS

STRATUS

NAME	HEIGHT/ALTITUDE	DESCRIPTION
CIRRUS	high, about 20,000 ft.	feathery, wispy, thin white sheets, like lace; made mostly of bits of ice; may produce snow or rain
ALTOSTRATUS	about 10,000 ft.	dense veils, thicker sheets
ALTOCUMULUS	about 10,000 ft.	rolls or layers of fluffy clouds
STRATUS	low/surface	low grey or white sheet of thick clouds; light drizzle or fog
NIMBOSTRATUS	surface to 6,000 ft.	grey sheets; rain or snow
STRATOCUMULUS	to 6,000 ft.	low, irregular mass
CUMULONIMBUS	to 20,000 ft. +	tall and anvil-shaped; dark clouds that often produce thunderstorms, lightning, rain, and/or hail; can also signal tornadoes
CUMULUS	to 6,000 ft.	low, fluffy, cauliflower-like; usually means good weather

Types of Precipitation and Storms

Cirrus and nimbostratus clouds are filled with moisture in the form of ice. Tiny ice crystals clump together to form snowflakes, and snow is produced. Sometimes the precipitation begins in the clouds as snow, but the ice crystals melt as they fall through warmer air and come down as rain. Because there are sometimes layers of both warm and cold air, precipitation can start as rain, then fall through a colder layer of air and turn into freezing rain or sleet.

Hail is produced when moisture travels up and down in a circle inside gigantic, tall, cumulonimbus clouds. As the balls of ice drop through the clouds, they pass through warmer air and pick up more moisture on their surfaces. Now, you remember that heated air rises. As the forming hailstones reach the heated air in the lower part of the cloud, the heated air carries the wet ice balls back up into a colder layer of air, where the collected moisture freezes on the surface of the ball of ice. The more the hail goes around in these up and down circles between cold air and warm, moist air, the more ice the hailstones collect. Sometimes the hailstones can become as large as golf balls, and when they fall, they cause great damage to cars, and especially crops in the field.

Can you think of any other weather system that involves clouds and air going in circles? If you said **tornados**, you were right. It's no surprise, then, that hail and tornados both come from cumulonimbus clouds. You know that rain and storms can be produced when warm and cold ocean currents meet; you also know that rain and storms can be produced when warm and cold air masses meet over land. When hot air and cold air masses "run into" each other, violent weather can be produced. These colliding air masses often produce cumulonimbus clouds, which can dump great quantities of rain or hail in a short time. These clouds can also produce severe thunderstorms, complete with lightning and even tornados.

Words to Know

Hail is precipitation in the form of ice, formed in thunderclouds.

Tornados are dangerous, whirling, funnel-shaped pillars of wind that reach down from cumulonimbus clouds to the ground.

Tornados are "born" from cumulonimbus clouds, and form a twisting, funnel-shaped cloud of wind that reaches from the cloud to the ground. These spinning winds, sometimes called "twisters," can do tremendous damage to buildings, crops, and anything else in their paths. Tornados have even destroyed whole towns.

These frightening storms generally form over large expanses of relatively flat land, like the Great Plains, and can form at any time of year. However, they are most common in the spring and fall, when there are more extremes of cold and warm air masses, waiting to collide.

Tornados can also form over the water. These rather unusual tornados are called tornadic **waterspouts**, and usually are fairly harmless. Sometimes, though, they pull water up into themselves, and then drop it onto the land in a great gush. There are a number of occurrences, including some in New South Wales; London, England; and Providence, Rhode Island, of waterspouts sucking water up from the oceans, and then dumping both water and flopping fish onto cities! (You may have heard the expression, "it's raining cats and dogs," but who would think of "raining fish"?)

Meteorologists and Storm Chasers

Knowing what kind of weather systems are coming our way is important to all of us. For students, knowing tomorrow's weather might help them prepare for school, for a snow day, a soccer game, or swimming at the beach. For farmers, being able to predict the weather can mean the difference between drowned

---

Some weather sayings:

These weather sayings are not always true, but in my experience, they are at least as good as any other method for predicting the weather. See if these work where you live. Sometimes these sayings only work in certain places, since they are based on observations.

Rain before seven, sun by eleven (morning, not night).

Red sky at night, sailors' delight; red sky in morning, sailors take warning.

Wind from the west we like the best;
Wind from the east we like the least;
Wind from the north, go not forth;
Wind from the south blows bait to the fish's mouth.

(This last part means it is a good time to go fishing.)

and rotted seed or young plants killed by frost, and healthy crops to harvest. For airplane pilots, knowledge of the weather helps them decide when and where it is safe to fly. For people who live in the path of a tornado or hurricane, accurate weather prediction can mean the difference between life and death.

People who study weather systems and predict weather are called **meteorologists**. You can see what an important job they have! People who may or may not be meteorologists, but have a keen interest in the weather, sometimes help meteorologists by observing and recording weather events near them. These "weather spotters" and "storm chasers" then report their findings to meteorologists, who can use this information to report the immediate weather even more accurately. Storm chasers drive or even fly near or into large storms like tornados and hurricanes. This practice can be very dangerous, as sometimes the chasers are caught in these violent storms.

Weather spotters usually report information that they see right in their own neighborhoods, calling the information in to a special contact meteorologist. Sometimes you can hear these reports coming in over the radio or television in the middle of a bad storm. Perhaps someday you would like to be a weather spotter for your community, or maybe even a storm chaser!

Words to Know

A tornadic **waterspout** is a tornado that forms over water.

Meteorologists are scientists who study and predict weather.

Climate

The term "weather" describes short-term changes; weather can change from minute to minute. Most of us have experienced days that had sunshine, some kind of precipitation, and wind all in the

same day, sometimes even changing from one to the other several times during the day.

Patterns of weather in a specific region over the long term are called **climate**. Climate doesn't change day by day, but can and does change over a period of years or centuries. While weather can change from day to day, we think of climate as staying the same. Desert regions are hot and dry most of the time; tropical regions are hot and moist most of the time; arctic regions are frigid most of the time. Temperate climates are pleasantly warm in the summer and not so terribly cold in the winter. If you are traveling to a tropical climate, you know that you don't need to pack a parka, ever. If you are traveling to the far north, don't even think about taking a bathing suit.

Yet even climate doesn't stay the same, but follows patterns of change. Climate changes occur naturally in God's plan; these changes can occur over seasons, years, centuries, thousands of years, and so on, so there really is no such thing as a "normal" climate.

Words to Know

Climate describes weather in a specific region, averaged over a period of years.

Where I live, in some years the rivers flood and all the crops are ruined. In other years there is drought when our wells run dry and the streams and ponds dry up. One year, we had green grass in January and some flowers began growing, and the very next year we had snow on the ground from the end of October to early May. Normal climate is the average of weather over several years, but each particular year is always just a bit different from the average.

Climates have changed drastically over centuries, and over periods of thousands of years. We know that in what are now desert climates, long ago there were lush forests, and some frozen climates were tropical long ago. During the Middle Ages, starting about one thousand years ago and lasting about three hundred years (at a time in history when there were no cars, trucks, or industry as we know it), large parts of the Earth went through a period of global warming. For those centuries, warm-weather crops grew where there is now ice much of the year. Our planet is now experiencing much cooler global temperatures than were experienced in the Middle Ages.

Measurements: English and Metric

Measuring Weather

Meteorologists, other scientists, and even your family, keep track of or record the weather on a daily basis. In your family, you might simply note the temperature. Others record wind speed, amount of rain, and a host of other things. Some countries record temperatures in Celsius, others in Fahrenheit. Which system do you use?

English and Metric Systems

Celsius temperatures set freezing at 0 degrees, and boiling at 100°. Fahrenheit sets freezing at 32°, and boiling at 212°. Which is better for measuring temperatures, Celsius or Fahrenheit? Both are fine, because both systems accurately measure cold and heat.

Just as different countries use different systems to measure temperature, countries also have different ways to measure distance. Some countries use the English system of inches, feet, and miles; others use the metric system of centimeters, meters, and kilometers.

Which is better, miles or kilometers, the English system or the metric system? Both are fine; they both accurately measure distance. I personally like to use miles because it is historic, but some people find kilometers easier because that is what they grew up using. Both systems are fine.

In the United States, miles are used officially, although kilometers are used by many people as well. In Canada, both miles and kilometers are used; kilometers are used on road signs, but many people speak and think in miles. Europeans tend to use kilometers as well, and the metric system is used in science more than the English system is.

English Measures of Length

English measurements are often based on simple everyday things from history. Inches, feet, yards, and all the rest are based variously on the lengths or widths of barleycorn; someone's thumb, hands, and feet; and rods. The mile was based on one thousand paces of a Roman soldier, dating back to the long-ago days of the Roman Empire, when Roman soldiers trod on English soil.

Units

barleycorn (this is still used for shoe sizes sometimes)
3 barleycorns to the **inch** (the basic unit, a blacksmith's thumb width)
4 inches in a **hand** (used to measure the height of horses)
12 inches in a **foot** (a bit large for a foot, but about right with boots on)
36 inches in a **yard** (about a pace, or large step)
5 ½ yards, or 15 ½ feet in a **rod** (I have heard this used by farmers for measuring fence lengths)
220 yards, or 40 rods,
    or 1/8 mile in a **furlong** (used in land surveying)
1,760 yards, or 8 furlongs, or
    5,280 feet in a **mile** (from the Latin *mille*, or 1,000 paces)
6 feet in a **fathom** (distance between left and right finger tips if a man holds his arms out; still used for measuring water depth, ropes, and cable lengths)

The Invention of the Metric System

The metric system was invented by rationalist scientists during the French Revolution more than two centuries ago.

The people who invented it were trying to come up with a purely scientific system based on reason alone, that did not include any cultural or human elements. They believed that measuring based on human things such as feet, hands, and inches, made the measurements too ordinary, and that measurements should only be based on pure reason. They followed René Descartes, who said, "I think, therefore I am."

These scientists were foolish because they were trying to use reason without God. They forgot that God is Truth, and that He is the most reasonable of all. They also forgot that we *are* human beings, made in the image of God, and that we humans are the ones who do the reasoning.

St. John Paul II said it best when he answered Descartes with, "I am, therefore I think." A person must *exist* before she or he can reason. Our humanity comes first. This seems so obvious that it is hard to believe that sometimes even clever people forget it!

The ungodly French Revolution was responsible for the deaths of many great martyrs, for when God is pushed out, sin and sorrow flow in. Yet, despite persecution and martyrdom by the rationalists, whole villages stayed loyal to Christianity, particularly in the Vendée region of France. To this day, the total number of heroic martyrs from the Vendée region is unknown.

Meters and Metrics

The metric system is based on the number ten. There are 10 millimeters in a centimeter; 10 centimeters in a decimeter; 10 decimeters in a meter; 10 meters in a decameter; 10 decameters in a hectometer; and 10 hectometers in a kilometer.

Metric Measures of Length

UNIT	NUMBER OF METERS
micrometer	1,000,000 in one meter, or 1,000th of a millimeter
millimeter	1,000 in one meter
centimeter	100 in one meter
decimeter	10 in one meter
meter	1 in one meter
decameter	10 meters
hectometer	100 meters
kilometer	1,000 meters

To convert miles to kilometers, just multiply the miles by 1.609. To convert kilometers to miles, multiply the kilometers by 0.621. To convert yards to meters, multiply yards by 0.9144. To go from meters to yards, multiply meters by 1.093.

Benefits to Each System

It is easy to learn the metric system, and it is easy to do math with the metric system. Because of this, many people consider metric to be a better system of measurement. Indeed, for many purposes, such as scientific measurements, it is better. Metric is the official system in Europe and Canada.

However, the English system is superior for other activities, and that is why it is still the official system in the United States. Even in Canada, which is officially supposed to be metric, most people use the English system every day simply because it makes more sense to do so.

For example, to make a perfect right angle, one can use the Pythagorean theorem to show that a triangle of sides 3, 4, and 5 units will always make a right angle. The numbers 3 + 4 + 5 add up to 12, the number of inches in a foot in the English system.

Using feet and inches makes the most sense for construction and carpentry. The English system is also more flexible. Using the metric system, numbers can be divided easily by 1, 2, 5, and 10. With the English system, numbers can be divided by 1, 2, 3, 4, 5, 10, and 12 easily.

The important thing is to know both systems and use the one that is most sensible for a given task. For science, this can mean the metric system, but for lots of other things English measurements make the most sense.

When we measure and study the vast Earth we live on, the atmosphere that towers above it, and complex weather systems that travel across great continents and oceans, we begin to understand how tiny we are in comparison to all of creation. Yet the One who created all this—and us, too—is also our *Father* who loves us, who is as close as our tabernacles, and our hearts. *Dominus Deus Sabaoth*, Lord God of all the Heavenly Armies, and *Abba*, Father!

Let's put what we've learned to work.

Roll Up Your Sleeves!

Please remember to read all of the activities, right to the end of the chapter, even if you are assigned only one or two activities. Information contained in the activities is also instructional, and part of your lessons!

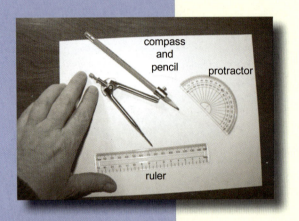

Activity #1

You will need a compass, protractor, ruler, pencil, and paper.

Mapping the Earth and Lines of Latitude

In this chapter, I stated that we were about halfway between the equator and North Pole. How do we measure where we are exactly? Our location between the equator and the poles is measured in degrees of latitude. Look at a map or globe for lines of latitude, which run east and west.

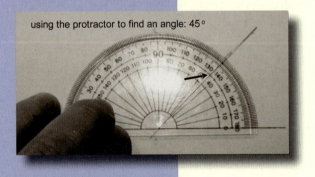

Behold and See 5 Student Workbook: Worksheets for Chapter 8 begin on page 59.

Now let's draw an example of how latitude works. First, draw a square, taking care to make it perfect, with each side 4 inches long. Find the center of the square by connecting the corners with diagonal lines (see Figure 1).

Next, we will learn how to use a compass to make our "map." A compass is a tool that has a point on one end and a pencil on the other. You place the point on a piece of paper, and then draw perfect circles and arcs with it. By measuring the distance between the pencil and the point, you can draw circles of the exact size you need.

Now use your compass to draw a circle inside the square. Place the pointed end of the compass exactly in the middle of the square, and make the circle 4 inches in diameter, so it just touches all four sides of the square. (See Figure 2.)

Set your compass to about two inches. Put your compass point on A, where the diagonal line crosses the circle, and draw a small arc above the square. Now do the same thing on B. The two arcs should cross each other. (See Figure 3.) Draw a straight line through the intersection of this arc and on through the center of the circle. You have just used geometry to divide the circle in half very precisely.

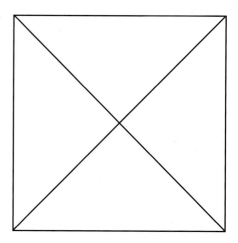

Figure 1.

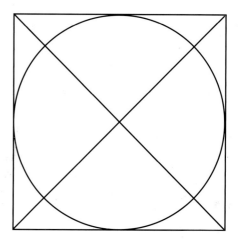

Figure 2.

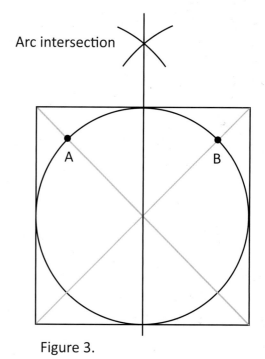

Figure 3.

165

Now use the protractor and make a mark on the circle 23.5° to the right of the center line. (23.5° is the amount the Earth's axis is tilted.) Draw a line from this mark directly through the center of the circle. Label the line 90° where it intersects the circle on the top and bottom. This is where the North and South Poles are. Now draw a line exactly perpendicular (90°) across this line, through the center to make the equator. Label this line 0° where it intersects the circle. (See Figure 4.) Each degree around the circle is a degree in latitude, and indicates how far north or south a place is from the equator.

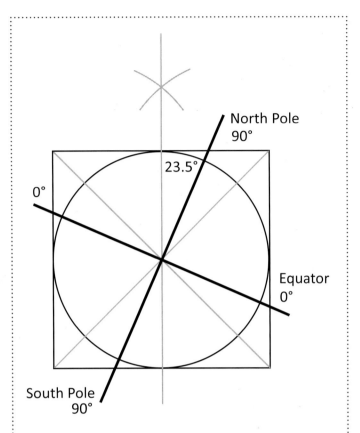

Figure 4.

Using the protractor, now mark the degrees on the edges of the circle from 0 to 90 by 10's. Mark them on both sides of the axis (the line through the poles) and join the marks to make lines across the Earth parallel to the equator. (See Figure 5.) These are the latitudes in the northern hemisphere. You just drew the exact measurements across the Earth the same way map makers do.

Where do you live? What is the latitude? Look it up on a map.

I live at 44.5° north latitude, which I put on the figure using a protractor.

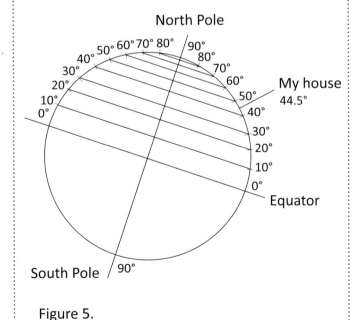

Figure 5.

166

Why is mapping important to science? There are many reasons. Remember that geologists are scientists who study the Earth? And marine biologists study animals that live in the oceans? What if these scientists heard about a great discovery somewhere on the Earth? If the scientists didn't know mapping, they might have a hard time marking the location of new discoveries, and finding what they want to observe! (It's handy to know if you want to go treasure-hunting, too.)

Activity #2

Have you ever thought that you might like to be a storm chaser, and track tornadoes, hurricanes, or typhoons? With this website, you don't need to own a car or even know how to drive.

www.stormpulse.com

Go to the website, and pick a storm to "track," or follow. Note where the storm is located, then return to the website in an hour or two and note where the storm has moved.

Has the storm intensified and gotten stronger, or subsided and grown weaker? How many states, provinces, or territories does the storm cover? Was the storm significant enough to be mentioned in the news? Did it cause damage? If the storm was a hurricane or cyclone, was it given an official name?

Activity #3

Get some paints, markers, or colored pencils. If you have paints, mix all the colors together. What color do you get? It will be black, or at least very dark brown. If you do this with all colors in exactly the same amount, you will get black paint. The material that absorbs and reflects light is called a pigment, and you see a color because the pigment absorbs every other color, and reflects only a single color back to your eyes. So a red apple reflects red light back to your eyes, but absorbs every other color.

So things have color because they absorb all light except that color.

If a pigment works by absorbing all the colors except the one you see, then what colors must be in ordinary sunlight? That's right, all the colors! After all, if there wasn't any red light in sunlight, then there wouldn't be any red light for the apple to reflect back to your eye. The same goes for all the other colors.

Activity #4

Get a piece of white paper and some pieces of colored paper, such as black, blue, red, green, and yellow. Put these in a sunny window for few minutes with the sun shining on them. Which is the warmest paper? Which is the coolest paper? How long does it take to notice which becomes warmer? The colored paper absorbs, or takes in, the light and the white paper reflects, or bounces back, the light. The reflectivity of a surface is called its **albedo**. (Shiny surfaces, like polished metal, are reflective for a different reason.)

White paper reflects every color, so this tells us that light made up of all colors together is white light. Why is the white paper cooler than the colored paper when they are both in the sun? Because the other colors of paper absorb some sunlight making them warm. But the white paper reflects all the sunlight so it doesn't warm up.

Remember these points:

1) All colors of *pigment* mixed together make black pigment which reflects no light.

2) All colors of *light* mixed together give white light.

Words to Know

Albedo is the capability of a surface to reflect light.

Activity #5

For a fun project, make some weather gauges and record the weather. Record hourly temperature, and the daily maximum, minimum, and mean (average) temperatures. You can get these from a newspaper or from a weather station as well, and compare them to the temperatures in your yard. Are they the same?

Check the temperature over your driveway or sidewalk, over some grass, bare soil, and under some trees or shrubs. How does the temperature differ in different places in your yard? These differences in temperature are called microclimates; animals choose the best microclimates to live in.

Temperature

Any thermometer will work for this. The best thing is to get a maximum/minimum thermometer, and write down the highest and lowest temperatures each day.

Rain gauge

This is simply a ruler in an open jar, to measure how much rain has fallen.

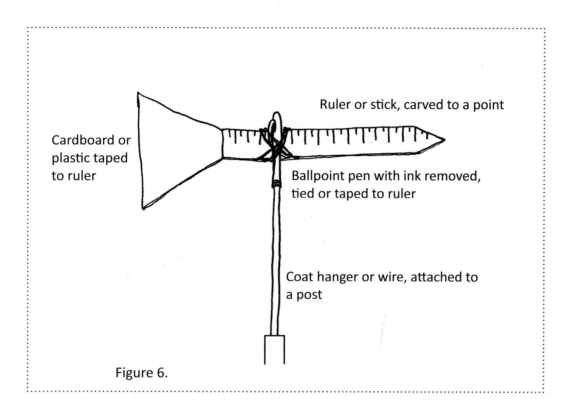

Figure 6.

Wind direction

This is a weather vane. You will need a post or pole, such as an old ski pole, and something that will rotate freely on it. The one in Figure 6 is very simple. You should be able to make a better one.

Barometer

If you can get a barometer, great, but do not spend a lot of money. Try to find a second-hand one. A barometer measures whether air is rising up, or falling down. Rising air creates a low pressure, and falling air creates a high pressure. Recall that rising air will produce clouds, and then rain, so a drop in pressure means rainy weather, and a rise in pressure means good weather ahead.

Chapter 9

The Earth and Its Composition

Thou art, O God, the life and light
Of all this wondrous world we see,
Its glow by day, its smile by night,
Are but reflections caught from Thee—
Where'er we turn, Thy glories shine,
And all things fair and bright are Thine.
— Anonymous

The Earth and Its Composition

Now that you know about the atmosphere above the Earth, and how it influences our weather on the surface, let's "dig deep" and find out what's underneath the Earth's surface!

To imagine what the Earth is like, think of a large ball of porridge or oatmeal that floats in space and is crusted over with sections of hardened oatmeal. The crust floats on the liquid oatmeal beneath. This helps us picture what the Earth is like. Because we live on the **crust**, we think it is solid and permanent, but in reality, it is a thin layer of hardened rock that floats on a ball of liquid and melted rock, called the **core**.

The Earth's crust is not smooth; it is pitted and dented in some spots. Think of a large, soft basketball that is no longer firm and round, because some of the air has gone out. If this basketball were left in the yard during a rainstorm, some of the dents or low spots on the ball would fill with rainwater. The "dents" or low parts of the Earth are full of water, too, and the high parts are more or less dry, with smaller "dents" of lakes and rivers of fresh water. The lowest sections are the oceans, and the high sections between the oceans are the continents. But remember that the oceans are not bottomless; they "float" on top of the Earth's crust.

Layers of the Earth

The Earth has layers just as the atmosphere does. The very center of the Earth is called the **inner core**, and it is a solid sphere of extremely hot iron. The iron inner core remains solid, even though it is so hot, because the enormous pressure of the surrounding liquid core prevents the iron from melting and becoming a liquid. The inner core rotates, similar to the way the Earth rotates, only slightly faster!

Words to Know

The Earth's **crust** is a layer of hardened rock that floats on a core of melted rock.

The Earth's **outer core**, located between the crust and inner core, is made of liquid iron.

The Earth's **inner core**, at the very center, is solid iron.

The **outer core** is liquid iron, and very hot as well. The fact that the outer core is liquid means it can rotate a bit more slowly than the inner, solid core, and a bit faster than the crust. This unusual motion produces electrical currents, which in turn produce the **magnetic field** that surrounds the Earth. Ninety percent of the magnetic field surrounding the Earth is made in the outer core.

Try this: Get a raw egg and spin it. Stop it, and then let it go again. If you do this, it will start to spin again, because the yolk is still spinning even though you stopped the egg shell from spinning. The yolk spins faster than the egg shell, just the way the Earth's core spins faster than the Earth's crust.

Beyond the outer core is the region called the **mantle**. The mantle is also made up of melted, or molten, rock. The lower mantle is dense and hot, and the upper mantle is cooler and less dense. The crust and the upper edge of the mantle together are called the **lithosphere**, which is about 43 miles (70 km) thick. We live on the crust of the lithosphere.

Words to Know

The Earth's **magnetic field** is a force field that protects the Earth from dangerous solar winds.

The Earth's **mantle**, made up of molten rock, is just below the crust.

The Earth's **lithosphere** contains the crust and the upper edge of the mantle.

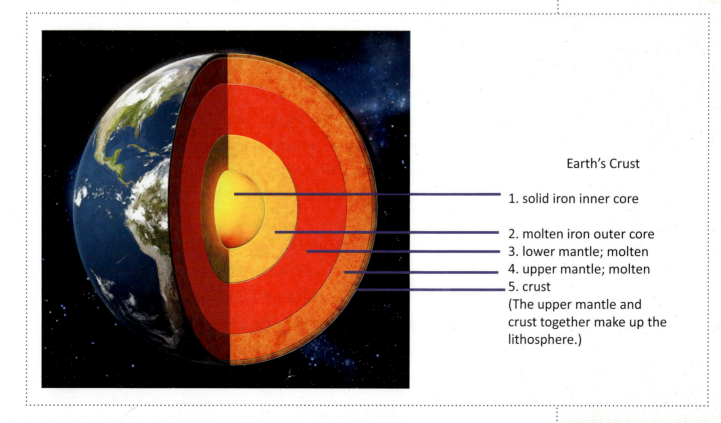

Earth's Crust

1. solid iron inner core
2. molten iron outer core
3. lower mantle; molten
4. upper mantle; molten
5. crust
(The upper mantle and crust together make up the lithosphere.)

Words to Know

Continental drift describes the spreading apart of the continents.

Magma is molten rock below the Earth's surface.

A **plate** is a large, moving piece of the Earth's crust.

Plate tectonics is the study of movements of sections of the Earth's crust.

The Earth's Crust

The crust of the continents is made out of granite, and the crust beneath the oceans is made out of basalt. Granite is less dense than basalt, so it floats on the basalt layer, which makes the continents. The granite continents can float higher or lower on the basalt layer, depending on their weight, so their height can change over time. To understand how this works, think of a log floating in the water. If you add some weight to the log, it floats a little lower in the water. When the weight is removed, the log bounces back to the surface.

The continents act the same way. When the weight of huge glaciers added to the weight of the land, the continents sank lower. Then, as the glaciers melted into the oceans, the land "bounced" back up, only very slowly. There is an island in James Bay, just along the southern shore of the Hudson Bay in the Arctic Ocean, where lots of Snow Geese and Canada Geese like to breed in the summer. It is called Akimiski Island (you can look it up on a map). This is a bounce-back island, and is only a few thousand years old. Most of Akimiski Island is swampy and low, but it is still rising.

Plate Tectonics

As the continents float on the basalt layer, they not only move higher and lower but they also move very, very slowly around on the surface of the Earth. (Of course, we cannot feel this motion because it happens so slowly.)

If you look at a map of the world, you will notice that the east coast of North America and South America seems to fit in a sloppy way into the west coast of Africa and Europe. This observation was first noticed by map makers and geographers in the early 1500's, and they started thinking about what could be the

cause. Then, in 1915, a geologist and meteorologist from Germany named Alfred Wegener suggested that the continents had been together at one time and had somehow drifted apart. (Do you see the discoveries that studying geography, geology, and meteorology can lead to? I wonder what *you* might discover!)

Wegener's idea that the continents were once all together as a single piece of land, but were somehow pulled apart and began to drift, was a good one. But Wegener could not explain how this might have happened, and many scientists rejected his theory. It was when scientists later began to map the ocean floor that they realized he might be right, and a possible explanation for **continental drift**, or the spreading apart of the continents, was presented.

In the 1960's, two scientists, Robert Dietz and Henry Hess, suggested that the ocean floor was splitting and spreading apart. They were pretty sure from their observations that melted **magma** from below the Earth's crust was filling in the crack, so that new ocean floor was being created as the separate pieces of the crust spread apart.

In other words, the crust of the Earth is "cracked" into several large pieces, which slowly move as they float on the liquid core. These floating, moving pieces fit together in a sort of Earth-size jigsaw puzzle! The term for this idea of moving pieces, or **plates**, is **plate tectonics**.

The Earth's Magnetic Field

Why is the Earth's magnetic field important? Because it protects the Earth from solar wind, a powerful electrical force that comes from the sun.

You probably know that every magnet has a north and south pole, and a magnetic field between those poles. The Earth is sort of like a huge bar magnet, with its own magnetic force-field. The Earth, as a gigantic magnet, also has magnetic north and south poles, not far from the regions that we call the North and South Poles. Magnetic force comes from the northern pole and invisibly flows around the entire planet to enter again at the southern pole. This strong, invisible magnetic force, constantly circling through the Earth and outward again, keeps the sun's radiation and electrical forces from striking our planet. God thinks of everything, doesn't He?

Words to Know

An **earthquake** is a violent shaking of the ground caused by movement of the Earth's plates.

Fault lines are places where the Earth's plates meet or crack.

The newest part of the Earth's crust is at the bottom of the Atlantic Ocean, up the center line between the Americas on one side of the Atlantic, and Europe and Africa on the other. The plates that those continents are on are separating along a "line" thousands of miles long, called the Mid-Atlantic Ridge. Because the ocean floor is getting wider, the Americas and Africa and Europe are being pushed farther apart. Of course, you can see the problem that arises: if the ocean floor is getting wider at one point on the globe, either the crust on the other side must buckle up, or it must be taken away.

The Earth's crust does both. In some places, the crust buckles up to form mountain ridges. In other places, especially the mountainous west coast of North and South America, the edge of the mountainous crust pushes down, down into the Earth's magma layer, and melts. The rock that was part of the mountains is sliding back down into the magma layer, to be melted once more into liquid rock.

The entire crust of the Earth is divided into plates that grow on one edge as fresh magma is added to them, and then slip under the plate

beside them along another edge. The slipping of one plate under or along another causes an **earthquake**, or a violent shaking of the Earth. Depending on how far below the surface of the Earth these quakes happen, and how close they are to populated areas, they sometimes aren't felt at all. And sometimes they are so strong that they cause tremendous damage. You may

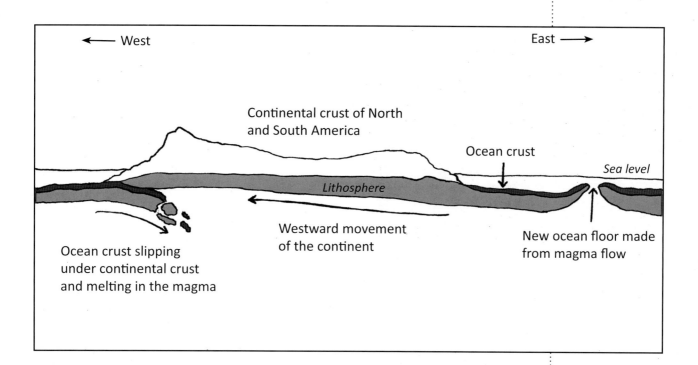

have heard of the famous quake that destroyed San Francisco, California in 1906. More recently, there have been devastating quakes in Haiti, China, Italy, and many other countries as well.

When the plates slip, it is not a smooth movement, as if the edges of the plates were greased. Rather, thick slabs of rock, often hundreds and hundreds of miles long, are sliding over other thick slabs of rock. As the plates slide, they often catch and stick. After years of pressure from the weight of the plates pushing, pushing against each other, they suddenly unstick and slip, sometimes violently. These points on plates where sudden slippages cause "tears" in the surface of the Earth are called **fault lines**, along which earthquakes can occur. Earthquakes are unpredictable, because we can never be sure when or how far plates will slip. But if you live near a major fault line, you can expect the Earth to shake now and then!

Earthquakes occur somewhere on the globe every day, but thankfully, most earthquakes cause little or no damage.

Volcanoes

Volcanoes and volcanic activity are often related to earthquakes. Volcanoes are formed when molten magma forces its way to the surface of the Earth, bursting a hole in the crust and flowing out to relieve the pressure. (Magma that spills out from volcanoes is called **lava**.) The weak parts of the crust give way, and the melted magma pours up and through the crust. As the magma cools, it builds up into tall cones over time. These tall cones are volcanic mountains, and are the typical **volcanoes** that erupt, or shoot out lava, every so often.

What would you think if you were outside playing soccer in your backyard one day, and a volcano started to grow in the middle of your backyard? Such things have happened!

In Mexico, in 1943, a farmer was working in his fields when he noticed a steaming crack in the ground. In about a week, enough magma had poured out of that crack that a cone taller than a five-story building was created, right there in that poor farmer's field! Within a year, the mountain had grown to more than 1,000 feet tall. With all the lava and ash burping out of the volcano, people in the area had to leave or get buried in ash and lava. But today, Mt. Paricutin is quiet.

Now, the edges of the Earth's plates are their weakest parts, and that's where the most volcanoes are found. The edge of the Pacific

Ocean, which is shaped in a sort of giant circle from the southern islands of Asia, up along the volcanic islands of Japan, across the Pacific to Alaska, and down the west coast of the Americas, has a lot of volcanoes. This big "circle" experiences so much volcanic and earthquake activity that it is sometimes called the **Ring of Fire**.

Within this Ring of Fire is Japan's most famous and beautiful volcanic mountain, Mt. Fuji. The state of Alaska has several active volcanoes, and the state of Washington experienced a terrible volcanic explosion in 1980 when Mt. St. Helens blew its top, burying forests—and some unfortunate people—in ash for miles and miles around. The enormous cloud of ash even reached into the stratosphere! Another volcanic eruption in Indonesia in 1815 poured so much ash high into the atmosphere that the ash circled the planet. Sunlight was significantly blocked, so that even crops did very poorly, and in some parts of the world, 1816 was called "the year without a summer."

While the Ring of Fire is one of the most active areas for volcanic activity, volcanic activity occurs elsewhere, too. One of the more famous eruptions in history happened in 79 A.D., when Mt. Vesuvius in Italy erupted. Hot gases and ash poured over the homes and people of Pompeii and Herculaneum, completely burying the cities. More recently, terrible eruptions have occurred in the Philippines, and a volcano growing under a glacier in Iceland hurled so much ash into the atmosphere that airplane traffic over much of Europe had to be cancelled for some time.

Think about this: if 70% of the Earth's crust is covered by oceans, and only 30% is land, what percentage of volcanic activity occurs on land? If you guessed about 30%, you were right. That means that about 70% of volcanic activity is under water!

Words to Know

Lava is molten rock that bursts from volcanoes.

Volcanoes are mountains that form when magma is pushed up onto the Earth's surface.

The **Ring of Fire** is a "circle" of volcanic activity surrounding the Pacific Ocean.

Words to Know

Igneous rock is rock that is formed by volcanic activity.

Sedimentary rock is formed from chemicals, sand or other sediment that packs together and turns into rock.

Metamorphic rock is rock that changed into another type of rock under heat or pressure.

Minerals are pure forms of chemical compounds, often in crystal form.

Now, underwater volcanoes have an important job, because they form more land. Some people mistakenly think that islands float on top of the ocean, but nothing could be further from the truth. Islands are actually the very top parts of mountains growing upward from the ocean floor. The Hawaiian Islands all grew from volcanoes that are still active, and erupt on a regular basis. The melted rock pours out, runs down the mountainside into the ocean, hits the cooler ocean water, and sends out great clouds of steam. The water cools the lava, and guess what it turns into? More rock, which means more island. If there were no volcanoes under Hawaii, there would be no Hawaii.

Rocks and Minerals

Because magma from volcanoes comes to the surface of the Earth and becomes part of the crust, much of our rock started with volcanoes. But there are three main categories of rocks, based on how they were made. These are **igneous**, **sedimentary**, and **metamorphic** rocks. **Minerals** are pure forms of some chemical compounds, and rock is a general term for mixtures of minerals.

Igneous Rock

Igneous rocks are made out of molten magma that hardens and cools. (Igneous means "fire-formed" in Latin.) Some igneous rock is black, hard, and bubbly, and really looks like something that was thrown out of a fire! Other volcanic rock is very light, like pumice, which is sort of a bubbly foam that hardened into rock. Pumice is a rock that can float, because it is so full of air bubbles. Another volcanic rock is obsidian, which is a very hard, almost glassy, black rock. Like glass, obsidian is very sharp, and the Native Americans were skilled at chipping obsidian to make sharp spear points and arrowheads for hunting.

Now, when molten magma cools down, different minerals in the magma can turn into beautiful crystals, or a sort of sparkly rock. When magma cools down quickly, the crystals are small; if magma cools slowly, the crystals are large. **Granite** is an igneous rock composed of quartz, feldspar, and mica crystals. Granite often has gold, silver, and other valuable minerals in it. (*Now* would you like to take up geology, the study of rocks?) The components of granite are sometimes very small and packed together, and at other times have large amounts of pure minerals.

Igneous rocks are the main components of the Earth's crust, and are the most common. The rock that forms the continents is largely granite, and as you remember, basalt is the rock that forms the crust beneath the oceans.

Sedimentary Rock

Sedimentary rocks are formed from chemicals and sands that are carried by water, wind, or ice, and then packed together and buried. Think of muddy rivers and creeks after a spring flood. The dirt or sand carried in the water is sediment that sinks to the bottom of the creek, or onto the land that the river floods. The sand or other sediments, often along with dead insects or plants, are then cemented together into hard rock. This happens either chemically, or under pressure from the sheer weight of all the layers of mud and sand. This is very similar to the way concrete is formed, and many sedimentary rocks look a bit like concrete.

If you have hard water, you will be familiar with scale, which is lime deposits on your plumbing pipes. This lime scale is a sedimentary rock, and is the cement that holds limestone together. Some sedimentary rock is made up of sand or small gravel or stones. Sandstone, siltstone, coal, shale, and **conglomerates**, or various types of small rocks "cemented" together to form new, larger rocks, are examples of sedimentary rock.

Words to Know

Granite is a type of igneous rock.

Conglomerates are a type of rock made up of smaller rocks "cemented" together.

Sedimentary rocks can include **organic matter,** which is decayed or decaying material that was once living. Sea shells, dead trees, leaves, and animals are all organic matter that can become sedimentary rock, built up over centuries and centuries and centuries, layer by layer, the pressure turning the lower layers to stone. Coal, which we use to produce electricity and to fuel and heat homes, businesses, and schools, is a sedimentary rock; so are limestone and shale.

Sedimentary rocks sometimes contain **fossils**, which are traces or imprints of long-ago plants or animals—especially animals with shells—that were buried in the mud and sands that turned into rock. Maybe you can find fossils near your home, in road cuts where sedimentary rock layers are exposed.

Metamorphic Rock

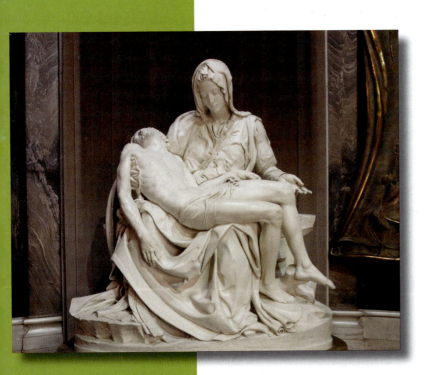

Michelangelo's Pietà

The third type of rock is called metamorphic rock. Metamorphic means that the rock was changed by heat and pressure into another kind of rock. The process is similar to cooking, because the structure of the material changes. The word metamorphic comes from a Greek word which means "to change"; it has the same root as metamorphosis, which you remember refers to insects changing from egg state to adult state.

Metamorphic rocks can be made out of igneous rocks or sedimentary rocks. For example, after centuries and centuries of pressure from layers of sediment above, limestone becomes lustrous marble that can be carved into beautiful statues, such as Michelangelo's Pietà. The sedimentary rock shale becomes slate; granite becomes gneiss; and sandstone becomes quartzite. Incredible pressure occurring along the edge of tectonic plates can produce metamorphic rocks as well.

Minerals

You remember that minerals are pure forms of chemical compounds, often in crystal form. Precious gems like diamonds, emeralds, and rubies are all minerals. Different minerals have different properties. Some are very **soluble**, or able to be dissolved, in water. One mineral that you are very familiar with, which is easily soluble, is salt! Stir a spoonful into a glass of hot water and watch it dissolve. Something that dissolves is soluble.

Minerals also have distinct colors, and different hardnesses. The hardest mineral is diamond, which has a hardness of 10. You cannot scratch a line in a diamond with your fingernail! The softest minerals include talc (from which talcum powder is made), asbestos, mica, and soft coal. These soft minerals, with hardnesses of 1 or 2, can easily be "dug into" with a fingernail. Calcite has a hardness of about 3 or 4; feldspar is 5 or 6; quartz is 7.

Notice that minerals can also fall into rock categories, such as igneous or sedimentary rock.

Fossils

Fossils are rocks that have insects, bones, plants, animals, or animal tracks or imprints in them. Some fossils contain the original animal or plant; others are just impressions made by the objects on the rock when it was becoming solid, or getting cemented together.

Many fossils are jumbled together, and so are ruined. But sometimes, if one is lucky, an intact or whole fossil can be found.

Words to Know

Organic matter refers to decayed or decaying material that was once living.

Soluble means "able to be dissolved."

Fossils are traces or imprints of plants and animals that existed long ago.

I usually find snail fossils, but have also found some interesting fossilized animal parts. Fossils can often be found in limestone, at least the ones I find are. Sometimes fossils are found in coal, too.

We usually think of fossils as being very old, but they are being formed even now under mud slides and river beds and anywhere that sediment is being deposited.

Do you recall from the chapter on plant defenses that some trees coat insects with sticky sap? If a tree with this sap falls into the water and sinks, the sap hardens and gradually becomes a beautiful mineral called amber. Lots of insects can be found in amber, perfectly preserved from thousands of years ago.

Some researchers specialize in finding fossil insects in the bottom of streams. These fossils are usually insect heads that might be hundreds of years old, mixed up in the sandy sediment. The deeper you dig, the older the insect heads are. This way scientists can find out what kind of insects were in a stream at different times in the past.

If you live near sedimentary rock deposits, you can collect fossils and minerals, too. I usually find fossils along stream edges where there are limestone layers.

Sand and Soil

Rocks and minerals can be carried by water, wind, and ice. When this happens, the rocks become rounded, smooth, and polished. This polishing action grinds the rough edges from rocks, and grinds up small rocks into even smaller pieces. Eventually, a layer of finely ground rock is formed, and this collects in different places. Water can grind down rocks by tumbling them together, and wind wears rock away by blasting it with dust, sand, and rain. Some plants, like lichen, can dissolve rock chemically.

When the rock particles become small, they are called sand. Shale ground down to powder is called clay. Larger particles are called silt. If any dead plants or animals get mixed up with the sand, clay, or silt, what do you think this is called? That's right! Topsoil! And, can you remember from the first chapter of this text, what mixes topsoil up so that plants can grow?

Right again, earthworms! As we reach the end of the school year, we are back where we began, only with a much better understanding of how everything around us fits together, and how miraculous it all is!

> Let's put what we've learned to work.

Roll Up Your Sleeves!

Please remember to read all of the activities, right to the end of the chapter, even if you are assigned only one or two activities. Information contained in the activities is also instructional, and part of your lessons!

Activity #1: Making fossils

You can make fossils in a way close to how they are made in nature. The basic idea is to cover a shell or leaf with soft material and make an imprint.

Use some modeling dough or salt dough for this. First, mix the dough:

- 2 cups flour
- ½ cup of salt
- ½ teaspoon of powdered alum
- ¾ cups of water (You can use coffee or tea instead of water to make it look more like a rock.)

Mix the dry material in a bowl, then add the water or coffee. Work it with your hands until it is like putty. Divide this into small balls, and push a ball of this over a leaf, snail, or shell so that the shape of the leaf goes onto the dough. Take the shells and leaves out of the dough and then place your fossils on an ungreased cookie sheet.

> **Behold and See 5 Student Workbook:**
> Worksheets for Chapter 9 begin on page 70.

188

Bake these "fossil cookies" for 30 minutes at 250°F in the oven. Then turn them over, and bake them another 90 minutes. Take them out and let them cool down. Now you have made fossils.

Here is another recipe that does not use the alum and doesn't need baking:

Mix ¼ cup of salt and ¼ cup of water, coffee, or tea with 1 cup of flour. Work this in a bowl. You can then form balls and make the fossils the same way as in the previous recipe. This should be left to harden for at least 24 hours, perhaps a bit longer.

It is a good idea to experiment with these recipes until you find the one that works best.

Do you notice something about these fossils you made? They are all perfect because you carefully formed them.

This is not how it is done in nature though. Nobody walked by and put some clay over a snail or leaf, and then let it harden or cooked it to make fossils! In nature, the material or clay (for which we substitute dough) and the fossil animal or plant gets mixed up in a jumble in the bottom of a stream or lake. Whole fossils in rock are rare, because they get broken and have missing parts. In nature, because fossils are formed in a random process, the fossils are rarely in nice, smooth, perfect layers.

Activity #2

Let's try to imitate nature's method of fossil-making with an experiment.

Make up a double batch of dough. Now, instead of making balls and imprinting, toss some fossil material directly into the mixing bowl. For example: snail shells, sea shells, small flowers, a few leaves, blades of grass, a worm, a fly, some small sticks, a grape or berry, a caterpillar if you have one, and a spider. Kill any live insects, worms, or spiders first by putting them in the freezer for a couple of days. Make sure you keep a list of what you included in the mix.

Roll out the mix into a large sheet, about ½ inch thick or even a bit thicker. Either bake it or let it dry, depending on which recipe you used.

When it is dry and hard, put on some safety glasses, put the slab in a cloth bag or pillowcase, and then smash it with a hammer, or over a stone or log. Smash it up into chunks, and then shake these so they are broken up.

Now, empty the bag, and see if you have any fossils. Identify what you can; keep count of each part. If you have some leaf bits in four separate pieces, then this counts as 4 fossils.

If you want to, make a graph to compare the number of each type of fossil material that went into the dough with the number that was counted after you made the fossils.

Which ones did you find?

Which ones were missing?

Which ones were hard to identify?

What do you think this says about the fossils that we find in nature? Obviously, soft bodied things do not fossilize very well, and it is rare to find good, whole fossils. Instead, you get lots of bits and pieces.

Now you can understand that whenever scientists or historians look at the past, or make conclusions using fossils, figuring out the truth based on bits and pieces can be tricky!

Appetite crystal in feldspar

Biotite micah crystal

Fossil coral limestone

Activity #3

Collecting rocks is fun! Make a habit of picking up unusual rocks wherever you travel. See if you can find samples of igneous, sedimentary, and metamorphic rock. A used field guide to rocks and minerals will help you identify your finds. You can keep your rocks in a box, and put a label on each one (liquid white-out works well), giving each one a number. Then, write down in your notebook where you found the rock, and when.

Activity #4

Let's make magnets and a compass to help us understand magnetic fields.

If you hang a magnet from a thread, it will point north and south, lining up its own poles with the Earth's North and South Poles. We can use this fact to make a compass from a magnetized sewing needle.

Get a magnet; a refrigerator magnet will do. Rub the sewing needle across the face of the magnet in *one direction*. This will magnetize the needle a small amount. (I did this 30 times, but 100 times will create a stronger magnetic field.) If you float the needle in a cup of water, it will point north and south.

The first time, I made the needle float in a jar lid by sticking it through a wooden matchstick. This worked fine.

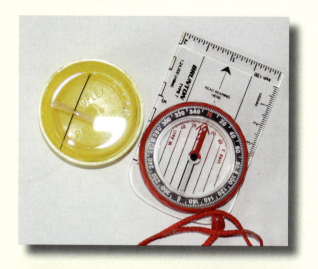

I then cut the bottom from a styrofoam coffee cup, taped the needle on this, and labelled it north, east, south, and west. Then I floated it in a bowl of water. The needle and cup bottom kept floating to the side of the bowl. To fix this, I cut a hole in the center of the cup bottom, and placed it over a spindle, which I made by sticking a needle upright in a block eraser. This held the bottom of the cup and needle so that it floated in the middle.

If you have a magnet, or something that you don't realize is magnetic such as a speaker or radio or large piece of steel nearby, your compass might not point north, but will instead point to the other magnet. Then you can find out what else in your house has a magnetic field, too!

10
Chapter

Genetics and Taxonomy

Order is heaven's first law.
—Alexander Pope

Genetics and Taxonomy

Words to Know

Genes are small pieces of information that form a sort of code on a chemical molecule called DNA.

DNA is the set of instructions, inside genes, for how the parts of living things are made.

Genetics is the study of genes and how living things are put together.

This year we have studied small things and big things in God's creation, from atoms and molecules and insects to rumbling volcanoes and the Earth and its atmosphere. Now let's return to study small things that are as close as your back yard. As you close out the school year and head into summer, you couldn't ask for a better time and season to collect specimens of plants and insects. Whether you're a "bug person" or a "plant person," there are many to choose from to start your collection.

Some of the greatest discoveries in biology have been found by scientists who collected simple weeds and insects. In fact, some— and I am one of these—would argue that the single greatest scientific discovery of the last 200 years is genetics.

In the nineteenth century, an Augustinian monk named Gregor Mendel grew a garden filled with pea plants. He planted pea seeds and cross-bred different colors and varieties. He observed the plants, and kept a careful record of the differences in the peas. His discoveries were the beginnings of the study of **genes**. Genes are small pieces of information that form a sort of code. They are stored on a chemical molecule called deoxyribonucleic acid, or **DNA** for short. The study of these genes and DNA is called **genetics**, or the science of how living things are put together.

DNA is the set of instructions inside each living thing that "tells" the plant or animal how to grow. DNA is found in each and every **cell**. Cells are the smallest whole pieces of a body part or plant part that can carry out the functions of that part. Parts like pea seeds or flowers, thumbs, hair color, eye color, feathers, skin, fur, bone shape, the attraction of some animals to bright lights—everything in fact that goes into making living animals or plants or bacteria is included in these instructions, or DNA.

The DNA code, made up of genes, is in each cell of all living things. And, if living things are preserved properly, this DNA code sometimes stays behind even after those things die. In this chapter, you will learn more about how you can collect, observe, and preserve living things and their code, much as Gregor Mendel did.

A good way to think about how DNA works is to imagine a book printed in a single line of words onto a long scroll of paper. Opening the scroll at different places would give different information. Imagine if such a book gave the instructions for making a car. Someone would only have to open the scroll at different places to find out how to make the engine, the doors, or windows.

Now, instead of words, imagine if the letters of each word were written in a mysterious code, such as the one below:

a is 10000001	j is 10000200	s is 10030000
b is 10000002	k is 10000300	t is 10040000
c is 10000003	l is 10000400	u is 10100000
d is 10000004	m is 10001000	v is 10200000
e is 10000010	n is 10002000	w is 10300000
f is 10000020	o is 10003000	x is 10400000
g is 10000030	p is 10004000	y is 11000000
h is 10000040	q is 10010000	z is 12000000
i is 10000100	r is 10020000	

Words to Know

A **cell** is the smallest whole piece of a body part or plant part that can carry out the functions of that part.

A **nucleus**, containing chromosomes, genes, and DNA, is the command center of every cell.

Words to Know

Chromosomes are twisted strands of DNA.

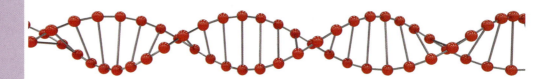

This code includes other things such as the number 44444444 to separate each word; 22222222 in front of a word to capitalize it; and 33333333 to insert a period.

If you knew the code, and found a scroll with this code written on it, you could read all sorts of secret messages. If your friends knew the code, too, you could write messages to them, and they would understand. DNA is a little like that: God wrote the code, and scientists are learning to read it.

The DNA code works like this: four different chemicals—adenine, thymine, guanine, and cytosine—make up the code parts. These chemicals are symbolized as A T G C (just like the numbers 1, 2, 3, and 4 in our code). Where the chemicals go tells what the "words" are, only instead of words and sentences, the A T G C codes are called genes. The genes are grouped into usable sections to make proteins and other materials, like bone, skin, blood, hair, or for birds, feathers. The strand of DNA is made up of long, twisted ladders of chemicals, so that A goes with T and C goes with G:

A—T

C—G

C—G

G—C

T—A

T—A

G—C

A—T

. . . and so on

To see how the DNA folds back on itself, get an elastic band, cut it open, and then twist it. Then, once it is twisted about seven times, bring your hands close together. The elastic will turn back on itself into a kind of twisted knot. This is exactly what the DNA does in each cell.

Imagine an elastic band about a mile long, and twisted into a long knot. A miniature version of this is in each cell. These twisted strands of DNA are called **chromosomes**, and people have 23 pairs of them in each cell. These chromosomes have our body's information written out in A, T, G, and C, with a special group to separate the genes, and with start and stop codes.

It is all so simple now that we know something about how genes, DNA, and chromosomes works. But before there were photographs of DNA or microscopes that could see chromosomes, nobody knew how parents passed down their characteristics to their children.

Gregor Mendel wanted to know how parent pea plants produced offspring that were the same as the parents. He had nowhere to begin except by keeping track of the characteristics of peas when he planted new seeds from parent plants. Now, Gregor Mendel didn't know he had discovered genes; it wasn't until more than a century had passed that genes were identified. But Mendel's careful research with peas laid the foundation for genetics. His discovery of how genes work happened because he was patient and kept good records in his notebook, as well as being a very clever scientist. Lots of important discoveries happen because collectors preserve specimens and keep good notes.

The Amazing Cell

Cells are remarkable things to study. To picture a cell, think of a five-story brick building, with the bricks as cells. The bricks, or cells, are very small in comparison to the whole structure, but without each brick, there would be no structure. Every living thing, plant and animal alike, is made up of these "bricks," or cells.

Of course, bricks are not the smallest things in a building, for they are made of smaller pieces of clay. It is that way with cells, too. Atoms, not cells, are the smallest "whole things" that can exist, and cells are made of atoms and molecules.

Inside each cell is a **nucleus**, the "command center" which contains the **chromosomes**, genes, and DNA that determine the function of the cell. A heart cell, for example, will never become a bone cell. Because of the genetic information and "orders" contained in the nucleus, the heart cell will not grow into any other kind of cell, not bone, or skin, or anything else. It was created to be a heart cell, and that's what it will be.

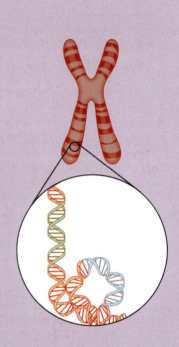

Chromosome

Now, you remember that a cell is the smallest whole piece of a body part that can carry out the functions of that part. Heart cells are good examples of that, for you know that your heart is a muscle that beats. But did you also know that if you take one, single, tiny heart cell away from the heart, that cell beats, too? And if a heart cell from one heart is added to cells from another heart, the different cells are soon all beating together in the same rhythm!

Isn't it amazing to think that you began your life as a single, tiny cell? From that very first moment of your life, all the chromosomes, genes, and DNA were packed into that cell that was YOU! The cell that was you already "knew" what color to make your eyes and hair and skin and a million more marvelous things. The tiny cell would not grow into a cow, or fly, or daffodil, for it was already *you*. Truly remarkable!

Now, most cells are too small to be seen without a microscope, but some are easily seen. An ostrich egg is a single cell; so are chicken eggs. And you have nerve cells that extend from the end of your backbone down to your feet.

Aren't cells amazing? I wonder Who designed them . . .

Scientific Information and the Names of Things

Sometimes **entomologists**, or scientists who study insects, in different parts of the world try to find ways of combating crop pests. When they discover new insects, they give a scientific name to the new insects they discover.

The newly-discovered specimen is placed in a museum and is called a "type specimen." Then the scientist writes the description and the new name in a scientific magazine called a journal. (Perhaps someday you will discover a new specimen and write about it!)

Other scientists sometimes make the same discovery in another part of the world, without knowing about the first discovery. Even though it's the same animal or plant, they don't know this, so they give it a new and different name, and preserve it in their museum. The scientists write about it in a different journal, in another language. Now the same animal or plant has two different scientific names, which can get very confusing. This happens quite often, and the only thing that stops the confusion is because specialized scientists called taxonomists (you remember taxonomy) travel the world looking at these specimens to compare them.

One of the flies I work with is a kind of fly called a blow fly. The blow fly lives on dead animals. The fly is bright green, and is called *Phaenicia sericata* in North America and *Lucilia sericata* in Europe and Asia. It used to be called *Lucilia caesar* by some entomologists, but there is another kind of green bottle fly in Europe whose scientific name is *Lucilia caesar*. The confusion came about because nobody compared our North American flies with the original museum examples from Europe to see if they were the same fly or not.

Words to Know

Entomologists are scientists who study insects.

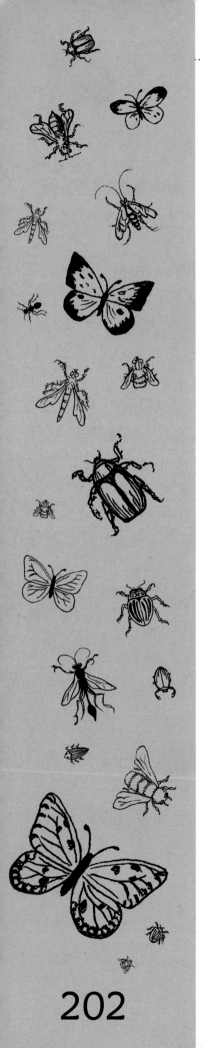

Why is this important? Some insects carry horrid diseases, and it is important to be able to know which ones are dangerous, and which ones are not. Lots of insects and plants are also being described in journals in other languages all over the world, so we can see how confusing it can get! To help stop some of the confusion, scientists give animals and plants Latin names, in addition to the animal or plant name in a local language. With Latin names used all around the world, there is at least a chance of getting things sorted out. If we only had common names in local languages, nobody would know what animals or plants they were talking about.

For example, we know that a certain little brown mosquito that likes to bite birds carries West Nile disease. It is called *Culex pipiens.* If mosquitoes only had common names like "Little Brown Mosquito," it would be impossible to know if the types of mosquitoes we have in North America are the same as the ones with the disease in Hungary, or France, or Bavaria. There are hundreds of different kinds of mosquitoes in the world, and most of them are brown. It would be a waste of money and time to try to stop West Nile disease by spraying brown mosquitoes in a marsh or swamp, because *Culex pipiens* mosquitoes live in eave troughs that catch rainwater from the roof!

Just to make matters even more exciting, we only know about 30% of the insects, and maybe 30% of the spiders in the world. We have just scratched the surface for lots of other animals like worms, mites, and ticks, and plants like weeds, fungus, mold, and algae. New species are being discovered all the time, so the world isn't going to run out of species to be discovered before you are ready to go exploring for them! (And sometimes the Latin name of the new plant or animal includes the name of the person who discovered it. Would you like to have a new fish, or flower, or perhaps a slimy slug named after you?)

We mentioned that scientific names are in Latin. There is a good reason for this, and it is why all really important words are in Latin and not English or French or some other language. Latin is a dead language, which means that it is no longer changing its meaning. English words change all the time. Just try to read Shakespeare, and see how the English language has changed in only a few hundred years. A computer used to mean a person who worked at a bank or factory, and sat at a table computing sums of numbers with a paper and pencil. There were sometimes hundreds of these people working in a large computing room, all keeping the records up to date. A computer is now a machine that was originally invented for computing sums, but today is used for almost everything else but addition.

In the Latin language, word meanings do not change, so if we read something written hundreds of years ago in Latin, we get the same meaning from it. Scientists, doctors, pharmacists, lawyers, legislators, philosophers, logicians, and mathematicians all use Latin, as well as anyone else who works with words where the exact meaning is very important.

Of course, the most important things, such as theology, Scripture, Church councils, and encyclicals, are all written in Latin for the same reason. It would be terrible if somebody tried to read a papal encyclical or document from a Church council written a thousand years ago, and had no way of knowing what the words actually meant! By using Latin as the language of Holy Mother Church, we can know very clearly the truths the Church has taught since Our Lord, Jesus Christ, founded it.

You remember that the scientific names of all living things are in Latin, and are classified to keep track of where animals and plants fit within their groupings. Knowing all this will make it easier for you to identify the specimens that you will collect.

Why Collect Specimens?

It is clear that collecting is a very important part of biology. It is how Linnaeus, who began the system of taxonomy, did his work, and how modern taxonomists, entomologists, and botanists still work today. Creatures like mammals and birds can be identified without being collected, but for small animals and plants like insects, spiders, or weeds, collecting is necessary.

There is an unfortunate modern tendency—seen in many of the new field guides and naturalist books—to downplay or deny the importance of collecting. **This is a mistake**. No photograph or written description preserves comprehensive information about an insect or plant, no matter how well done they are.

Collections provide:

1. a reusable source of data for ongoing studies linked to a particular time and place;

2. a verifiable specimen for re-identification and a historical record.

Collected specimens become available for future researchers. For example, much of the research that has identified phoretic species of mites has been based on museum and private collections from the past. Without new collections, this important research will disappear in the future!

Collecting takes skill, and the only way to learn it is to start! As with most skills, practice is required to produce useful specimens. Poorly prepared specimens with incomplete notes are useless.

You can collect weeds like Linnaeus did, or insects, or leaves, or sea shells, or fossils, bark, twigs, or anything! The main thing is to keep a record of the information in your notebook, and to keep the collection organized so that the data are preserved.

No matter what you choose to collect this summer, or how you choose to observe and preserve your specimens, you are now on your way to becoming a scientist.

Have a wonderful summer!

Roll Up Your Sleeves!

Let's put what we've learned to work.

Behold and See 5 Student Workbook: Worksheets for Chapter 10 begin on page 77.

Please remember to read all of the activities, right to the end of the chapter, even if you are assigned only one or two activities. Information contained in the activities is also instructional, and part of your lessons!

Activity #1

Let's do an exercise that helps us understand how DNA works as a code to tell living things how they are to grow. Using the code on the next page, see if you can decode the following messages:

What does this say?

Sentence 1: 22222222100000401000001010000400100004001000300033333333

Hint: separate out into individual codes, 8 characters long. You should be able to figure out how many letters there are without even decoding it.

Here is a harder one:

Sentence 2: 2222222210003000100200001000000144444444100040001002000010003000444444441000200010003000100000021000010010030000333333333

Now use this code to write your name in your notebook (remember to include the capital letter code and space code).

Notice that this code is written with only four different numbers, not counting zeros, which could be blanks. In fact, we can change the code so that it is written out with letters instead of numbers, so that 1 is A, 2 is B, 3 is C, 4 is D, and 0 is a blank space.

206

Code

a is 10000001	j is 10000200	s is 10030000
b is 10000002	k is 10000300	t is 10040000
c is 10000003	l is 10000400	u is 10100000
d is 10000004	m is 10001000	v is 10200000
e is 10000010	n is 10002000	w is 10300000
f is 10000020	o is 10003000	x is 10400000
g is 10000030	p is 10004000	y is 11000000
h is 10000040	q is 10010000	z is 12000000
i is 10000100	r is 10020000	

Capital letter is 22222222; period is 33333333; space between words is 44444444.

Can you write out Sentence 2 (previous page) in letters and spaces instead of numbers? Here is how it starts:

BBBBBBBBA _ _ _ C _ _ _ _ A _ _ B _ _ _ _ _ A _ _ _ _ _ _ ADDDDDDDD . . .

Here is a trick that we could use to make this code harder to crack: for every letter, put in a different letter, so that for B, put in A, and for A put in B, and for C put in D and for D put in C. Then we could copy this code using a code, and copy the copy to get back to the original.

Words to Know

An **herbarium** is a collection of plants.

Activity #2: General guidelines for plant collecting

A collection of plants is called an **herbarium**. The basic idea is to collect and press wild plants, and then glue them to white cardboard. These can then be placed in a binder.

Project instructions

Collect a minimum of 12 different plants, ideally about 20. Or you may wish to collect leaves instead. It is a good idea to focus your collection around a theme. For example, garden plants or weeds, or woodland plants or weeds.

Carry a small notebook and pencil with you at all times. When you find a specimen, note the date, time of day, location, and type of habitat. You will be tempted to ignore this advice, but resist this temptation and *always use your notebook*.

Be prepared to collect anywhere. Bring at least a rudimentary kit wherever you go, including a pocket knife (with parental permission), small shovel, and box to prevent damage to specimens.

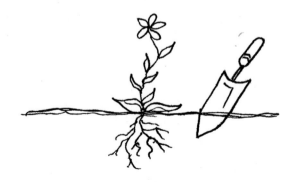

Collecting places

Look in all types of habitats. Look in ditches, along roadsides, gravel drives, woods, fields, along fences, anywhere that wild plants grow. Note whether the site has been recently disturbed (i.e. evidence of construction or plowing). Make sure you include all parts of the plant: flower (if present), stem, leaves, and roots. Collect seeds in an envelope and include them in the collection.

Pressing

The plants you collect should be pressed between newspapers and alternate layers of corrugated cardboard. Place a small board on the top of the pile. The whole set can then be weighted down under rocks, logs, or books. Or you can tie the bundle between boards with a rope or strap and tighten this as needed. Check your plant specimens each day to prevent them from becoming stuck to the newspapers. Waxed paper can be used between plants for final drying once most of the moisture has been removed. The entire process will take about a week if done indoors.

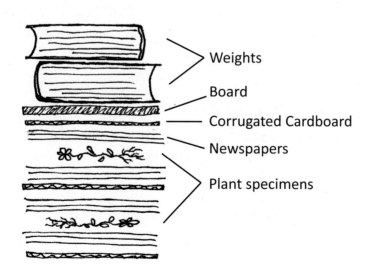

Gluing

Dried plants should be mounted using carpenter's wood glue or taped to herbarium paper. Coat some waxed paper with the glue. Then gently drop specimens into position on the paper. Do not shift specimens once placed; they will be ruined and the glue will remain as an unsightly mess. Taping may be easier if you use good-quality, clear tape carefully placed.

Labels

Labels are simply card stock cut to standardized sizes, about 10 x 10 cm (4 x 4 inches).

For writing the labels, India Ink is traditionally the best. Or, you can use fine black (not blue or red) ink. I am partial to labels written neatly in pencil.

The following information should be included:

1) Identification, including Family, Genus, and species, if possible. For this you will need a field guide of common wild plants or weeds for your region. These are not expensive when purchased used.

2) Collecting data, including collector's name (yours), place, and date—this label is *the most important*. Anyone can figure out what the plant is, even if you get its name wrong, but only you can provide the collecting information.

3) In what kind of habitat was the plant found? In a field, ditch, lawn, or woods?

Good hunting!

Activity #3: Collecting Insects: Entomology Collecting Guide

Project instructions

Collect a minimum of 12 different insects, ideally about 20. Collect several specimens from each species to see if they are slightly different. Also include male and females of the same species. You may wish to provide an overview of insects associated within a specific region, or insects that have certain biological features in common, such as pond insects, or night-flying insects attracted to light.

Carry a small notebook and pencil with you at all times. When you find a specimen, note the date, time of day, location, and type of habitat. You will be tempted to ignore this advice, but resist this temptation and *always use your notebook*.

To display your collection, you can use a cigar box, a shoe box, or better yet, a box from a hobby store used for displaying such things.

Collecting

Look everywhere! Insects are found in soil, under rocks and bark, in compost piles, on dead animals and abandoned nests, in flowers, weeds, shrubs, and water—both still and running. A very good insect net can be made using either mosquito netting or other material (e.g. "bridal veil") attached to the frame of an old badminton racquet. Some collectors dye their nets green, but I have not found that this helps. Rather, the large dark object seems to scare the insects away, particularly butterflies.

Use your net to sweep across the tops of weeds and flowering plants in parks, fields, gardens, and along fence rows. Even if you cannot see any insects before you sweep your net, you will be surprised that a few sweeps of the net will catch some that you did not see.

Words to Know

The **thorax** is the middle part of an insect's body.

There are as many specialized ways to collect as there are entomologists! For example, you may wish to try sugaring for moths. This involves painting a mixture of beer and molasses onto a tree trunk or fence post during the late afternoon. Then, in the evening or early night, check the trunk for any moths that have arrived. Light traps also work well, as does collecting beetles and moths attracted to outside lights at night.

Killing

The killing fluid used is ethyl acetate, which is the main ingredient in most inexpensive nail-polish removers. (Indeed, I prefer to use nail-polish remover as the killing fluid rather than more expensive chemicals.) Insects killed using ethyl acetate stay relaxed for spreading for about 24 hours.

With parental supervision and sufficient ventilation, set the insects you have caught in a wide-mouthed jar and place some killing fluid inside on some cotton or other absorbent material. You can immobilize butterflies or dragonflies before killing them by pinching them sharply on the sides of the **thorax**, or middle part of the insect's body, between your thumb and forefinger. (Do not use this method with moths, as it will only crush them.)

Try not to get any fluid on the insect; it will mat scales (if your specimen is a moth or butterfly) or damage some colors. A separate jar should be used for moths or butterflies to prevent other insects from being covered in their scales.

Make sure you keep insects in the killing jar for a few hours after they stop moving. It can be irksome to have an insect "return to life" late at night after being mounted on a spreading board.

Pinning

Pinning preserves insects for further study. Ideally, you should get some insect pins for this (get these from an online collecting or biological supply store). I will assume you have the right kind of insect pins, but if you cannot find pins, then you can glue dried insects to cards the size of postage stamps. (This is how beetles used to be collected.)

Push the pin through the insect from the top. Pins are placed off center, through the right side, so that the left side at least is undamaged. Many insects are identified by features that occur along the mid-line, or axis of symmetry, so it is important not to damage this area.

The top of the insect should be 10 mm or ½ inch from the head of the pin, with labels spaced below for easy reading. Obviously, large bodied insects must be placed higher on the pin. (The most versatile pin size is #3, but other thicknesses are available.)

Spreading

After pinning, an insect should be spread, wings and legs and antennae as needed, to appear lifelike, or at least so that features are not obstructed by curled legs. Use a styrofoam board for spreading. With parental supervision, use a knife to cut out a groove for the bodies of butterflies and dragonflies so that their wings can be spread out.

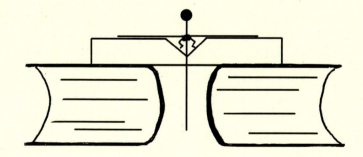

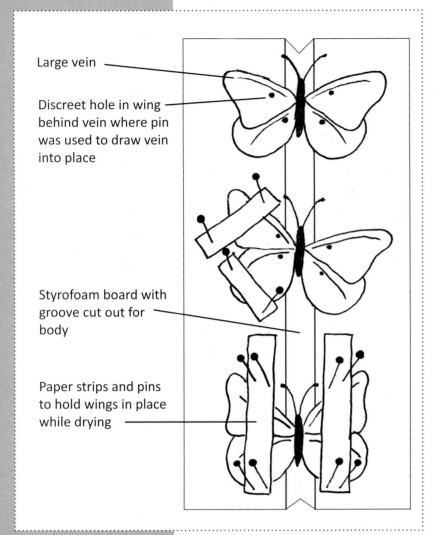

- Large vein
- Discreet hole in wing behind vein where pin was used to draw vein into place
- Styrofoam board with groove cut out for body
- Paper strips and pins to hold wings in place while drying

Spreading can be daunting at first, but the process is actually very simple. First, pin the butterfly. Then set it so that the wings are adjacent to the top of the foam board. You can use a pin to draw the wings into position, by inserting a pin directly behind the large vein running parallel with the top edge of the wing. Then using strips of paper, hold the wings in their final position as they are brought into place.

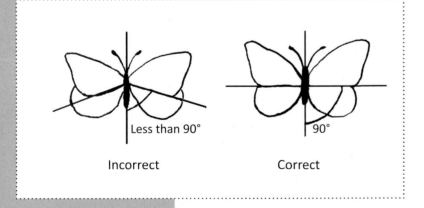

Less than 90° — Incorrect

90° — Correct

Butterflies and moths are mounted so that the hind edge of the front wing is perpendicular to the body. Dragonflies and other large winged insects are mounted with the wings perpendicular to the body. (If you would prefer not to pin, these insects can be kept in clear envelopes instead.)

Grasshopper with one wing spread

Bee with wings at 45° angles

Wasps and bees are often mounted with wings set at 45° from the body. (This varies; I prefer 90°.) Some grasshoppers are mounted with one side of the wings out and the other closed; it's your choice.

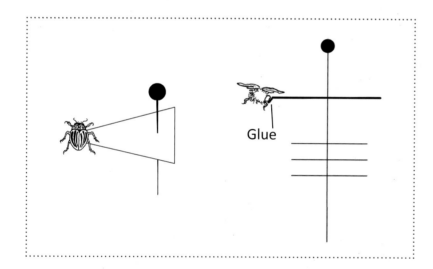

Small insects can be mounted on paper points and glued with white glue to the pointed part of the triangular cards. These are then mounted on #3 pins.

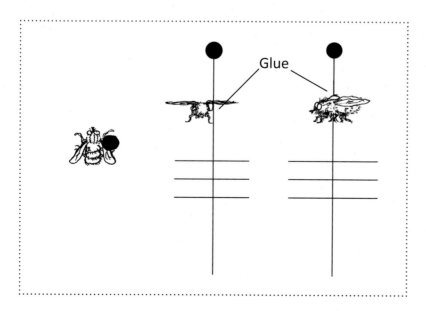

Smaller Diptera (flies) can be glued directly to the sides of #3 pins without points. You may use clear nail polish to glue small insects, or a water-soluble white glue, or wood glue. For all these mounts keep the pins to one side, ideally the right.

215

Soft-bodied insects such as caterpillars can be preserved in rubbing alcohol (ethanol) in small vials.

Labels

Labels are simply card stock cut to standardized sizes, 16 mm by 8 mm (¾ inch by ½ inch) or so. For labels, black ink is the best. Or, you can use pencil. Computer-printed labels are fine.

Three labels are necessary:

1) Identification, including Family, Genus, and species, if possible.

2) Collecting data, including collector's name (yours), place, and date—this label is *the most important*. Anyone can figure out what the insect is, even if you get its name wrong, but only you can provide the collecting information.

3) Biological or ecological information of note, such as caught in water, during the day, at night under a light, sweeping in an open field, under bark, etc.

For specimens mounted in vials filled with rubbing alcohol (ethanol), pencilled labels should be placed directly *in* the specimen vials. On no account should specimens be number-coded to a separate information page. The whole idea behind labels is that the information is kept *with* the insect.

Make sure that labels are at a 90° angle to the pins, with the pins through the center of the labels. Pins should not go through lettering.

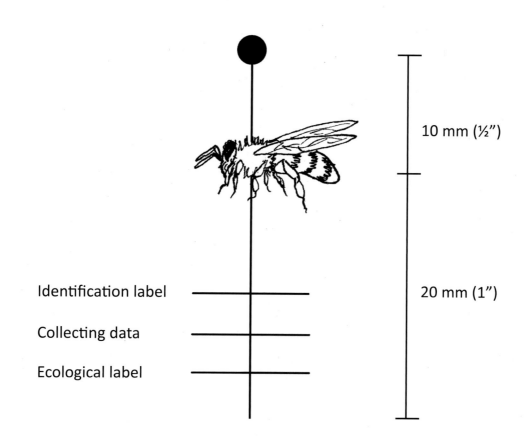

Arrangement of Collection

Group your insects together by Order. Label each order with a label held in place with small mapping pins or other short brass pins (sewing pins will do).

Carpet beetles, also called dermestid beetles, are everywhere, eating the dried skin and hairs between the cracks of your floor and furniture. These beetles will destroy any collection unless repulsed by moth balls or naphthalene flakes. **Note that mothballs and naphthalene flakes are now considered a health risk, and their use is not recommended any longer**. However, I do not know of any other way to protect insects from being destroyed by dermestids! If you wish to disregard this health warning, make a small cardboard box for the mothballs and keep this in place with pins within the collecting box. Every entomologist does this, but be sure to get your parents' permission.

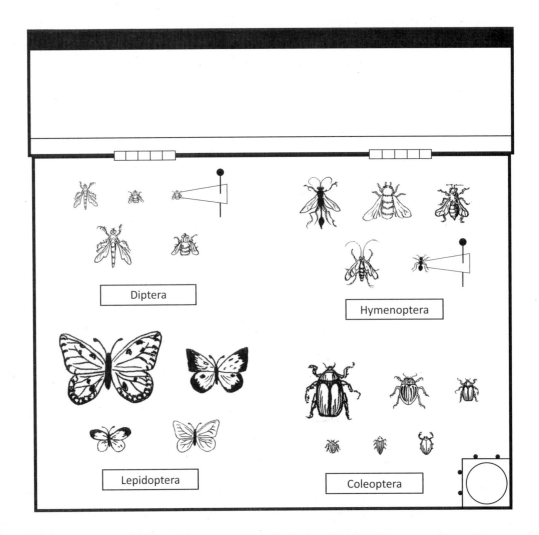

Extra Tips

Winged insects can be placed in envelopes before being killed in a killing jar; this prevents them from breaking their wings. This also allows more insects to be killed at one time. Similarly, robust insects can be placed in paper tubes when caught. Write any information on the tube or envelope itself. This way, you can collect even when you do not have all of your equipment on hand.

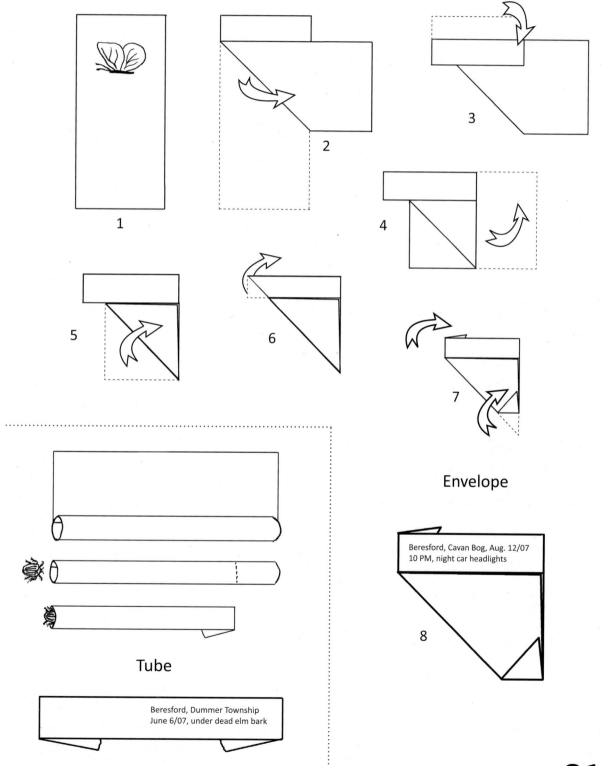

Indexed Glossary

albedo (al-BEE-doh): capability of a surface to reflect light—p. 169

altitude (AL-tuh-tood): measurement of distance from the surface of the Earth into the atmosphere—p. 146

anther (ANN-thur): flower part that produces pollen—p. 36

artery (AHR-tuhr-ee): blood vessel that carries blood from the heart to the rest of the body—p. 87

atmosphere (AT-muhs-fear): layers of air and gases that surround and blanket the Earth—p. 140

atom (AT-tuhm): the basic building block of the universe, and the smallest possible "whole piece" that a thing can be broken down into—p. 60

atrium (AY-tree-uhm): upper chamber of the heart—p. 90

axis (AX-uhss): imaginary center "line" around which the Earth rotates—p. 138

base (bayss): bottom of a food web, where the tiniest plants produce food from sunlight—p. 41

bioaccumulation (buy-oh-uh-kyum-you-LAY-shun): the increasing concentration of something (like chemicals) in animal populations—p. 46

biochemistry (by-oh-KEHM-uh-stree): study of chemical composition and reactions in living things—p. 70

biology (by-AWL-oh-gee): study of living things—p. 2

bonds (bahndz): join atoms together to form molecules—p. 72

botany (BOT-uh-nee): study of plants—p. 2

carbohydrates (kahr-boh-HYE-draytz): type of sugar produced mostly by plants, and made of carbon, hydrogen, and oxygen—p. 71

carbon cycle (KAHR-buhn SYE-kuhl): a system in which carbon moves through a cycle from the air to plants to animals and eventually back to the air to be used again and again—p. 80

carnivore (KAHR-nuh-vohr): creature that eats only meat—p. 41

cell (SELL): smallest whole piece of a body part or plant part that can carry out the functions of that part—p. 197

cellulose (SELL-you-lohs): plant fiber, made of a type of glucose—p. 79

characteristic (care-ek-tur-ISS-tick): special difference that makes a plant or animal unique, so that it is not exactly like any other plant or animal in its species—p. 29

chemical equation (KEHM-uh-kuhl ee-KWAY-shun): "recipe" for how chemicals are put together to form a material or cause a chemical action—p. 76

chitin (KYTE-uhn): a tough outer coating that is found on many insects and is made of a type of glucose—p. 79

chlorophyll (KLOHR-uh-fill): pigment that causes green color in plants; used for photosynthesis—p. 70

chromosomes (KROME-oh-sohmz): twisted strands of DNA—p. 198

circulatory system (SIR-kew-luh-toh-ree SIS-tuhm): system of the body that circulates oxygen and nutrients from the heart through the body—p. 86

classification (class-uh-fuh-KAY-shun): scientific system that groups things into classes with other like things—p. 9

climate (KLY-muht): weather in a specific region, averaged over a period of years—p. 158

compete (kuhm-PETE): to do your best to win something that someone else also wants—p. 114

condense (kuhn-DEHNSS): to turn from a gaseous state to a liquid state—p. 148

conglomerates (kuhn-GLAW-mehr-uhtz): type of rock made up of smaller rocks "cemented" together—p. 183

conservation of energy (kahn-sir-VAY-shun uv EHN-uhr-gee): scientific law that says energy can be neither created nor destroyed—p. 62

contest competition (KAHN-test kahm-puh-TI-shun): when two creatures compete but only one wins—p. 115

continental drift (kahn-tuh-NEN-tuhl drift): spreading apart of the continents—p. 176

core, inner: solid iron core at the very center of the Earth—p. 174

core, outer: liquid iron surrounding the solid inner core of the Earth—p. 174

crust, Earth's: layer of hardened rock that floats on a core of melted rock—p. 174

crustacean (kruhs-STAY-shun): animal whose skeleton is made from chitin and is outside its body (exoskeleton), similar to a shell—p. 121

cyclone (SAHY-klohn): severe storm that forms over the ocean—p. 152

decompose (dee-cuhm-POHZ): to rot or break down into nutrients—p. 44

deduction (dee-DUCK-shun): a type of thinking that uses general ideas or facts to explain a specific case—p. 100

defense mechanism (dee-FENCE MEH-kuhn-izzuhm): way that a body or plant protects itself—p. 122

DNA: abbreviation for deoxyribonucleic acid; the set of instructions, inside genes, for how the parts of living things are made—p. 196

earthquake (UHRTH-kwayk): violent shaking of the ground caused by movement of the Earth's plates—p. 178

element (ELL-uh-muhnt): chemical or substance that cannot be broken down into simpler chemicals—p. 72

emphysema (em-fuh-SEE-muh): a disease that destroys the lungs—p. 92

entomologists (ehn-tuh-MOLL-oh-jists): scientists who study insects—p. 201

equator (ee-KWAY-tuhr): dividing line that separates the Earth into two hemispheres, north and south—p. 138

equinox (EH-kwuh-knocks): day when daytime and nighttime are equal—p. 143

exhale (ECKS-hay-uhl): to breathe out—p. 87

evaporate (ee-VAPP-uh-rayt): to turn into a vapor, or gas—p. 148

facilitation (fuh-sill-uh-TAY-shun): a working relationship between creatures that helps both creatures—p. 22

fault lines (FAHLT lynz): lines along which the Earth's plates meet—p. 178

filament (FIL-uh-muhnt): flower part that holds up the anther—p. 36

food web: system in which sunlight and both tiny and large plants and animals, work together so everything in the cycle is nourished and fed—p. 41

fossils (FAW-suhlz): traces or imprints of plants and animals that existed long ago—p. 185

fungus (FUN-guhs): type of plant that is often harmful to other living things—p. 7

genes (jeenz): small pieces of information that form a code on a chemical molecule called DNA—p. 196

genetics (juh-NET-icks): study of genes and how living things are put together—p. 196

glacier (GLAY-shur): an extremely thick sheet of ice—p. 5

glucose (GLUE-kohs): type of sugar, made of carbon, hydrogen, and oxygen atoms—p. 71

granite (GRAH-nuht): type of igneous rock—p. 183

grub (gruhb): worm-like, larval stage of an immature beetle—p. 20

guerilla strategy (guh-RILL-uh STRAH-tuh-gee): when plants send out runners that start and connect several plants across an area—p. 124

habitat (HA-buh-tat): place or environment where something lives—p. 8

hail (HAY-uhl): precipitation in the form of ice, formed in thunderclouds—p. 155

hemisphere (HEM-i-sfeer): half of the Earth's sphere; the Northern Hemisphere is above the equator, and the Southern Hemisphere is below the equator—p. 139

hemoglobin (HEE-moh-gloh-buhn): pigment in blood that carries oxygen through our bodies—p. 89

herbarium (uhr-BARE-ee-um): collection of plants—p. 208

herbivore (UHR-buh-vohr): creature that eats only plants—p. 41

host (hoest): animal or plant that a parasite lives and feeds on—p. 24

human physiology (HUE-muhn fizz-ee-AWL-oh-gee): study of the cells, organs, and functions of the human body—p. 86

hurricane (HER-uh-kayn): severe storm that forms over the ocean—p. 152

hybrid (HYE-bruhd): cross between two older types of plant or animal that creates a new type or breed—p. 29

hypothesis (hye-PAW-thu-suhs): an educated guess about why something happens based on what we already know and can observe—p. 98

Ice Age: period of time when ice covered much of the continents—p. 5

igneous rock (IHG-nee-uhs RAHK): rock formed by volcanic activity—p. 182

immune (ih-MYUNE): able to "fight off" harmful germs or substances—p. 48

immunity (uh-MYOUN-uh-tee): ability of the body to "fight off" harmful germs or substances—p. 48

immunizations (ihm-you-nuh-ZAY-shunz): "shots" that keep people and animals safe from certain diseases—p. 48

induction (ihn-DUCK-shun): a type of thinking that examines many experiences to figure out what might happen most of the time—p. 100

inhale (ihn-HAY-uhl): to breathe in—p. 87

inherit (in-HAIR-uht): to receive characteristics passed down by a parent to its offspring—p. 29

invertebrate (in-VUHR-tuh-brayt): animal without a backbone—p. 10

larva (LAHR-vuh): worm-like, immature form of an insect—p. 11

latitude (LA-tuh-tood): measurement of distance north or south of the equator—p. 140

lava (LAH-vuh): molten rock that bursts from volcanoes—p. 181

lithosphere, Earth's (LIH-thus-fear): the crust and the upper edge of the Earth's mantle—p. 175

logic (LAW-jick): special way of thinking or reasoning—p. 98

lungs (luhngz): organs that fill with air when we breathe in, and send away carbon dioxide when we breathe out—p. 87

magma (MAG-muh): molten rock below the Earth's surface—p. 176

magnetic field, Earth's (mag-NEH-tick fee-uhld): a magnetic force field that protects Earth from dangerous solar winds—p. 175

manipulate (muh-NIP-you-layt): a sneaky way of making something do things that it doesn't want or plan to do—p. 122

mantle, Earth's (MAN-tuhl): molten rock, just below the Earth's crust—p. 175

masting (MASS-sting): cycle by which plants grow fewer seeds for a time and then more seeds in following years—p. 127

mesosphere (MAY-sohs-fear): third layer of the atmosphere from the Earth's surface—p. 145

metamorphic rock (meht-uh-MOHR-fik RAHK): rock changed into another type of rock under heat or pressure—p. 182

metamorphosis (met-uh-MOHR-fuh-sis): process of going through different, changing forms to reach adulthood—p. 20-21

meteorologist (mee-tee-ohr-AWL-oh-juhst): person who studies and predicts weather—p. 157

minerals (MIHN-uh-ruhlz): pure forms of chemical compounds, often in crystal form—p. 182

molecule (MOHL-uh-kewuhl): very tiny substance, made up of one or more kinds of atoms—p. 60

molt (mohlt): to shed skin, primarily in insects—p. 66

nucleus (NEW-klee-uhs): command center of every cell, containing chromosomes, genes, and DNA—p. 197

omnivore (OHM-nuh-vohr): creature that eats both meat and plants—p. 41

opinion (uh-PIHN-yun): a belief or feeling about something that is not a fact—p. 44

orbit, Earth's (OHR-buht): "path" the Earth takes as it travels around the sun—p. 139

organic matter (ohr-GAHN-ik matter): decayed or decaying material that was once living—p. 185

ovary (OH-vuh-ree): part of a flower that becomes fruit—p. 36

ovule (AH-view-uhl): part of a flower that becomes seed—p. 36

parasite (PAIR-uh-syte): creature that lives on, or feeds off of, other living things—p. 11

parasitoid (PAIR-uh-sit-oyd): somewhat like a cross between a predator and a parasite—p. 24

pesticide (PEST-uh-side): chemical used to kill insect pests—p. 8

phalanx strategy (FAY-lanks STRA-tuh-gee): when plants grow thickly together to crowd out competitors—p. 124

phoresy (FOHR-uh-see): a working relationship in which one creature "hitches a ride" on another—p. 22

photosynthesis (foh-toh-SIN-thu-sihs): process by which plants use sunlight to turn carbon dioxide and water into sugar with the help of chlorophyll—p. 70

plate (PLAYT): large, moving piece of the Earth's crust—p. 176

plate tectonics (PLAYT tehk-TAHN-icks): study of movements of sections of the Earth's crust—p. 176

pollen (PAW-luhn): part of the flower that makes flowers fertile, or able to produce fruit—p. 35

pollination (pohl-uh-NAY-shun): process of spreading pollen to flowers to produce fruit and seeds—p. 22

precipitation (pree-sip-uh-TAY-shun): water, in one form or another, falling on the Earth—p. 152-153

predator (PREH-duh-tore): animal that hunts for and eats other animals—p. 11

prediction (pree-DICK-shun): what we think might happen in the future if our hypothesis is correct—p. 98

prey (pray): creatures that are hunted and eaten by predators—p. 11

primary producer: plants and animals near the bottom of the food web—p. 42

pupa (PEW-puh): cocoon in which immature insects grow into adults—p. 11

pupate (PEW-payt): to finish growing into an adult insect inside a cocoon-like structure—p. 20

reproduce (ree-proh-DOOS): to produce offspring—p. 119

respiration (rehs-pehr-AY-shun): process of using oxygen to release energy—p. 76

Ring of Fire: a "circle" of volcanic activity surrounding the Pacific Ocean—p. 181

runners (RUHN-nerz): a kind of sideways stem that grows along the ground away from a parent plant; a new baby plant takes root at the end of the runner—p. 125

scramble competition (SKRAM-buhl kahm-puh-TIH-shun): when two or more creatures compete to get the best or most of something—p. 115

sedimentary rock (sehd-uh-MEHN-tree RAHK): rock formed by chemicals, sands, or other sediment that packs together and turn to rock—p. 182

sepal (SEE-puhl): flower part that protects the bud—p. 36

solar radiation (SO-luhr ray-dee-AY-shun): invisible rays or waves of energy, some harmful, emitted by the sun—p. 146

solstices (SOHL-stuh-siz): longest and shortest days of the year—p. 142-143

soluble (SAWL-you-buhl): able to be dissolved—p. 185

species (SPEE-sees): group of animals or plants of the same kind—p. 5

spores (SPORZ): similar to tiny, airborne seeds—p. 7

stigma (STEEG-muh): sticky part of the flower that catches pollen—p. 35

stratosphere (STRAT-uhs-fear): second layer of the atmosphere from the Earth's surface—p. 145

style (stile): flower part that supports the stigma—p. 36

T. gondii (tee GONE-dye): a type of parasite—p. 119

taxonomy (tax-ON-uh-mee): scientific name for classification—p. 9

thermosphere (THURM-uhs-fear): fourth layer of the atmosphere from the Earth's surface—p. 145

thorax (THO-racks): the middle part of an insect's body—p. 212

tornado (tohr-NAY-doh): dangerous, whirling, funnel-shaped pillar of wind reaching down from cumulonimbus clouds to the ground—p. 155

trichomes (TRY-combs): bristly hairs that grow on and help defend some plants—p. 126

trophic level (TROH-fick LEH-vuhl): a level in a food web—p. 42

troposphere (TROH-pohs-fear): atmospheric layer closest to the Earth—p. 144

typhoon (tahy-FOON): severe storm that forms over the ocean—p. 152

vapor (VAY-pohr): another word for gas; water vapor is water in its gaseous form—p. 144

variability (vehr-ee-uh-BILL-uh-tee): differences within a group or species—p. 29

vein (vayn): blood vessel that carries "used" blood back to the heart—p. 87

ventricle (VEHN-truh-kuhl): lower chamber of the heart—p. 90

vertebrate (VER-tuh-brayt): animal with a backbone—p. 10

volcanoes (vawl-KAY-nohz): mountains formed when magma is pushed up onto the surface of the Earth—p. 181

water cycle (WAH-tuhr SYE-kuhl): cycle of evaporation and condensation that provides a never-ending supply of water to the Earth—p. 148-149

waterspout (WAH-tuhr spout): tornado that forms over water—p. 157

zoology (zoh-OL-uh-jee): study of animals—p. 2

Supply List

CHAPTER 1
- pail or large bowl
- shovel
- ruler
- small board

CHAPTER 2
- bird feeder (optional)
- bird seed for many types of birds
- two large flowers
- cutting board
- knife

CHAPTER 3
- animals/insects to research (see options on pg. 66)
- container and food for your animals

CHAPTER 5
- 2 beef or lamb hearts
- knife
- cutting board
- camera (optional)

CHAPTER 7
- 9 small pots with soil
- radish seeds
- ruler
- 1 cup of earthworms or mealworms, or 4 hardboiled eggs
- large square of plywood or cardboard (2 ft x 2 ft)
- *Optional:* birdhouse activity, see pg. 135

CHAPTER 8
- compass
- protractor
- ruler
- pencil and paper
- paints, markers, or colored pencils
- white paper
- 4–5 pieces of paper of different colors

Temperature project:
- thermometer

Rain gauge project:
- ruler, jar

Weather vane project:
- post or pole, ballpoint pen, ruler, cardboard or plastic

Barometer project:
- barometer

CHAPTER 9
- modeling or salt dough, see pgs. 188-189 for recipes
- leaves, shells, flowers, grass, bugs, sticks, etc.
- safety glasses
- cloth bag or pillowcase
- hammer
- unusual rocks
- field guide to rocks and minerals
- box for collecting rocks
- labels or liquid white-out
- magnet
- sewing needle
- cup of water
- styrofoam cup
- block eraser

CHAPTER 10
Plant and insect collections may be completed after Chapter 10 or enjoyed as summer projects.

Plant Collection:
- 12-20 different plants or leaves
- pocket knife
- small shovel
- box to carry specimens
- envelopes for seeds

Pressing:
- newspapers, cardboard, small board, waxed paper
- wood glue and waxed paper, or tape and herbarium paper
- card stock paper
- field guide of wild plants for your region

Insect Collection:
- 12-20 different insects
- cigar, shoe, or collecting box
- insect net or mosquito netting and an old racquet
- nail-polish remover
- 1–2 wide-mouthed jars and lids
- a few cotton balls
- #3 insect pins, or card stock paper and white glue
- styrofoam board
- strips of paper
- rubbing alcohol in small vials (optional)
- card stock paper
- push pins or sewing pins
- mothballs or naphthalene flakes (see health warning on pg. 218)

Needed for All Chapters: *small notebook, pencil and colored pencils, magnifying glass (optional)*

Image Credits

Illustrations by AnneMarie Johnson. Illustration on pg. 135 and photos on pgs. 20, 22, 25, 37, 92, 110, 121, 164, 189, 192, 193, 213 © David Beresford; pg. 91 LifeART (Image © 2004 Lippincott Williams & Wilkins. All rights reserved.); cover © Kane Skennar/Digital Vision/Getty Images; cover, pg. i © Gorilla/Shutterstock; cover, pg. i © JinYoung Lee/Shutterstock; cover, pg. i © Martin Fischer/Shutterstock; cover, pg. i © Andreas Gradin/Shutterstock; pg. iii © Cathy Keifer/Shutterstock; pg. iv © Andreas Gradin/Shutterstock; pg. v © oksana.perkins/Shutterstock; pg. v © sjgh/Shutterstock; pg. v © Gorilla/Shutterstock; pg. v © JinYoung Lee/Shutterstock; pg. vi © Adrian Britten/Shutterstock; pg. 2 © Adrian Britten/Shutterstock; pg. 2 © LVV/Shutterstock; pg. 2 © Vulkanette/Shutterstock; pg. 2 © Galushko Sergey Alekseewisch/Shutterstock; pg. 4 © Lenkadan/Shutterstock; pg. 4 © Pakhnyuschcha/Shutterstock; pg. 5 © Antoine Beyeler/Shutterstock; pg. 6 © Teo Dominguez/Shutterstock; pg. 7 © Constant/Shutterstock; pg. 9 © Becky Sheridan/Shutterstock; pg. 9 © Olgysha/Shutterstock; pg. 9 © Arthur van der Kooij/Shutterstock; pg. 11 © Ronnie Howard/Shutterstock; pg. 12 © Mika Heittola/Shutterstock; pg. 12 © oxbeast1210/Shutterstock; pg. 13 © Andrei Nekrassov/Shutterstock; pg. 18 © ephotographer/Shutterstock; pg. 20 © ephotographer/Shutterstock; pg. 20 © fotosav/Shutterstock; pg. 21 © Alexander M. Omelko/Shutterstock; pg. 23 © Gorilla/Shutterstock; pg. 24 © Julien/Shutterstock; pg. 29 © Timur Kulgarin/Shutterstock; pg. 34 © Delmas Lehman/Shutterstock; pg. 36 © Daniel Tay/Shutterstock; pg. 38 © syzx/Shutterstock; pg. 40 © syzx/Shutterstock; pg. 41 © oksana.perkins/Shutterstock; pg. 43 © Andreas Gradin/Shutterstock; pg. 44 © zahradales/Shutterstock; pg. 45 © Cathy Keifer/Shutterstock; pg. 46 © FloridaStock/Shutterstock; pg. 46 ©FloridaStock/Shutterstock; pg. 49 ©Ermes/Shutterstock; pg. 52 ©Mark Herreid/Shutterstock; pg. 55 ©Elena Elisseeva/Shutterstock; pg. 56 ©Henrik Larsson/Shutterstock; pg. 57 ©FLariviere/Shutterstock; pg. 59 ©RoJo Images/Shutterstock; pg. 60 ©africa924/Shutterstock; pg. 60 ©africa924/Shutterstock; pg. 61 ©dofmaster/Shutterstock; pg. 62 ©Morgan Lane Photography/Shutterstock; pg. 65 © Mitch Aunger/Shutterstock; pg. 66 ©werg/Shutterstock; pg. 67 © PetrP/Shutterstock; pg. 68 © Mr Doomits/Shutterstock; pg. 70 © Mr Doomits/Shutterstock; pg. 70 ©mashe/Shutterstock; pg. 71 ©abrakadabra/Shutterstock; pg. 76 © Vaclav Volrab/Shutterstock; pg. 78 ©Ivan Montero Martinez/Shutterstock; pg. 84 © Benjamin Loo/Shutterstock; pg. 86 © Benjamin Loo/Shutterstock; pg. 87 ©Sebastian Kaulitzki/Shutterstock; pg. 93 © Igumnova Irina/Shutterstock; pg. 93 ©Jacek Chabraszewski/Shutterstock; pg. 94 ©TonyWear/Shutterstock; pg. 96 © Steinhagen Artur/Shutterstock; pg. 98 © Steinhagen Artur/Shutterstock; pg. 99 ©Tischenko Irina/Shutterstock; pg. 101 ©Mike Rogal/Shutterstock; pg. 102 ©Olaf Speier/Shutterstock; pg. 104 © Elena Elisseeva/Shutterstock; pg. 112 © JinYoung Lee/Shutterstock; pg. 114 © JinYoung Lee/Shutterstock; pg. 116 ©Stargazer/Shutterstock; pg. 117 ©afitz/Shutterstock; pg. 117 ©SNEHIT/Shutterstock; pg. 118 ©Magdalena Bujak/Shutterstock; pg. 120 ©Gertjan Hooijer/Shutterstock; pg. 122 ©Vasaleks/Shutterstock; pg. 123 ©Poznukhov Yuriy/Shutterstock; pg. 125 ©D & K Kucharscy/Shutterstock; pg. 126 ©Pontus Edenberg/Shutterstock; pg. 127 ©Geanina Bechea/Shutterstock; pg. 130 © Frank Chan Jr./Shutterstock; pg. 131 © Levent Konuk/Shutterstock; pg. 131 © Matt Ragen/Shutterstock; pg. 131 © voylodyon/Shutterstock; pg. 133 © 11018386/Shutterstock; pg. 134 © Steve Byland/Shutterstock; pg. 136 © Martin Fischer/Shutterstock; pg. 138 © Martin Fischer/Shutterstock; pg. 140 © Valery Potapova/Shutterstock; pg. 142 © saied shahin kiya/Shutterstock; pg. 142 © Tatiana Popova/Shutterstock; pg. 142 © Tatiana Popova/Shutterstock; pg. 142 © Tatiana Popova/Shutterstock; pg. 145 © oorka/Shutterstock; pg. 146 © Tyler Boyes/Shutterstock; pg. 147 © vitor costa/Shutterstock; pg. 150 © clarusvisus/Shutterstock; pg. 152 © photobank.kiev.ua/Shutterstock; pg. 153 © Lars Christensen/Shutterstock; pg. 153 © RAFAI FABRYKIEWICZ/Shutterstock; pg. 153 © Reeed/Shutterstock; pg. 153 © Irene Teesalu/Shutterstock; pg. 154 © mihalec/Shutterstock; pg. 154 © paul prescott/Shutterstock; pg. 154 © Chin Kit Sen/Shutterstock; pg. 154 © Ustyuzhanin Andrey Anatolyevitch/Shutterstock; pg. 157 © Dark o/Shutterstock; pg. 158 © STILLFX /Shutterstock; pg. 158 © Fred Hendriks/Shutterstock; pg. 158 © Dainis Derics/Shutterstock; pg. 161 © IgorGolovniov/Shutterstock; pg. 163 © Edwin Verin/Shutterstock; pg. 167 © Mike Phillips/Shutterstock; pg. 168 © Antoine Beyeler/Shutterstock; pg. 170 © Joy Brown/Shutterstock; pg. 172 © Chee-Onn Leong/Shutterstock; pg. 174 © Chee-Onn Leong/Shutterstock; pg. 175 © Andrea Danti/Shutterstock; pg. 177 © IgorGolovniov/Shutterstock; pg. 178 © iBird/Shutterstock; pg. 180 ©juliengrondin/Shutterstock; pg. 183 ©Don Bendickson/Shutterstock; pg. 183 ©icetray/Shutterstock; pg. 184 ©Eugene Mogilnikov/Shutterstock; pg. 186 ©markrhiggins/Shutterstock; pg. 187 ©D & K Kucharscy/Shutterstock; pg. 190 ©AleZanIT/Shutterstock; pg. 194 © Royik Yevgen/Shutterstock; pg. 196 © Royik Yevgen/Shutterstock; pg. 198 ©Aspect3D/Shutterstock; pg. 199 ©Brendan Howard/Shutterstock; pg. 200 ©Blamb/Shutterstock; pg. 201 ©Eric Isselée/Shutterstock; pg. 203 ©wjarek/Shutterstock; pg. 204 ©Brendan Howard/Shutterstock; pg. 205, 225 ©Morgan Lane Photography/Shutterstock; back cover: © oksana.perkins/Shutterstock.com; back cover: © ephotographer/Shutterstock; back cover: © oorka/Shutterstock.com; back cover: © Poznukhov Yuriy/Shutterstock.com; back cover: © markrhiggins/Shutterstock.com; back cover: © Valerie Potapova/Shutterstock.com